Why Dissection?

Why Dissection?
Animal Use in Education

Lynette A. Hart, Mary W. Wood, and
Benjamin L. Hart

GREENWOOD PRESS
Westport, Connecticut • London

Library of Congress Cataloging-in-Publication Data

Hart, Lynette A.
    Why dissection? : animal use in education / by Lynette A. Hart, Mary W. Wood, and Benjamin
L. Hart.
        p. cm.
    Includes bibliographical references and index.
    ISBN 978–0–313–32390–4 (alk. paper)
    1. Animal experimentation—Moral and ethical aspects.   2. Biology—Study and teaching.
3. Laboratory animals.   4. Dissection.   I. Wood, Mary W.   II. Hart, Benjamin L.   III. Title.
    HV4712.H37   2008
    179′.4—dc22   2007039199

British Library Cataloguing in Publication Data is available.

Library of Congress Catalog Card Number: 2007039199
ISBN-13: 978–0–313–32390–4

First published in 2008

Greenwood Press, 88 Post Road West, Westport, CT 06881
An imprint of Greenwood Publishing Group, Inc.
www.greenwood.com

Printed in the United States of America

The paper used in this book complies with the
Permanent Paper Standard issued by the National
Information Standards Organization (Z39.48–1984).

10 9 8 7 6 5 4 3 2 1

Dedicated to
The Many Teachers
Who Inspire and Inform Their Students
With the Wonders of Biology

Contents

Preface   ix

1. Today's Biology Classroom Crisis   1

2. Early History of Dissection: Controversy and Advances   17

3. Controversy in the Development of Science Education and
   Dissection   35

4. The Context for Dissection: Educational Testing   53

5. National versus State Educational Standards: Whither
   Dissection?   71

6. Legislation and Regulations Related to Using Animals and
   Dissection in Teaching   93

7. Empowering Teachers to Find Substitutes for Dissection   111

8. Students and the Culture of Dissection   133

9. The Animals Used in Teaching   151

10. New Teaching Resources in the Computer Age   165

11. Locating Teaching Resources and Research Literature on
    Alternatives   183

References   199

Index   217

Preface

Having a long-standing interest in studying and teaching about animal and human biology we were intrigued by the prospect of the engaging topic of this book when we were approached by Greenwood Press several years ago. Our combined experience in biology totals over one hundred years, during a period in which uses of animals in education have been transformed, especially within the field of advanced biology education where we have worked. Working within veterinary medicine has offered us a relatively unique perspective for experiencing the transformation of teaching methods during the past couple decades.

CURRENT STATUS OF DISSECTION IN PRECOLLEGE EDUCATION

Apart from the rich historical literature, we continue to be surprised at the scarcity of information on dissection, especially since it is such a pervasive aspect of secondary school biology education. Most of the published discussions on dissection are initiated by humane societies or animal welfare organizations advocating to eliminate dissection. These organizations provide lists of alternative teaching resources, offer assistance to students not wishing to perform dissection, and sponsor legislation providing protection to students against being required to participate in dissection. But in the professional education literature dealing with secondary schools, the topic of dissection has languished. To us it seems there is little consideration of the perspectives of teachers, students, and parents regarding dissection. Nor is there much substantive consideration of the use of animals for dissection. More importantly, the technological advances in imaging and computer interactivity are just beginning to be systematically employed for teaching precollege biology.

It seems surprising to us that dissection still continues in intermediate and high schools, much as it was fifty or more years ago, even though it fails to appear in discussions of educational standards and frameworks. Thus, we have set out to develop a book that would examine dissection from a range of perspectives, to help make sense of how we have arrived at our current practice and why it has been so slow to change. Readers will gain views of dissection representing each of the various constituencies related to the practice of dissection. We provide resources in the tables and in a companion Web site to facilitate further exploration of the topics as desired by readers: http://www.vetmed.ucdavis.edu/Animal_Alternatives/appendices.html.

During the past decade, our opportunities to talk with precollege science teachers in workshops and focus groups have informed this endeavor. The book is directed to all those teachers in secondary schools who are especially interested in advancing science education and furthering the scientific engagement of students, as well as those with concern regarding uses of animals for dissection in precollege. We have particular concern for teachers who enthusiastically seek to inspire their students to be excited about science, even as their available budgets and teaching materials shrink.

We are surrounded at the UC Davis School of Veterinary Medicine by bright and creative minds working to enhance the teaching of our veterinary students. Inspiring breakthroughs continually create more efficient learning opportunities based on reusable teaching resources. Bringing information technology to veterinary education through the increased use of software complemented with plastinated specimens and various anatomical molds, freeze-dried preparations, and professionally dissected prosections, is fostering an economical and very effective revolution in veterinary medical education.

Other universities have been phasing out their courses in comparative anatomy due to lack of materials, a growing cost of using animal specimens for dissection, and sometimes even a lack of expertise in comparative anatomy. Most alternative teaching resources developed in recent years have been spearheaded by a strong philosophical commitment to reduce "consumptive" uses of animals for education. Advocacy for abolishing uses of animals for dissection, and discussions among students and faculty concerning these uses, have stimulated improved access to teaching resources. What has been missing has been a serious consideration of the reasons for dissection continuing in precollege despite its termination in medical professional schools.

This book is the most complete and comprehensive analysis of the use of animals in precollege dissection. This practice is influenced by historical and philosophical traditions. Ramping up requirements in reading and mathematics to some extent is diminishing the resources remaining for science. Studying the various body systems fits within a specified area of the science curriculum as defined in educational standards and frameworks, yet no specific guidance is offered pertaining to dissection. Availability of teaching resources such as specimens, models, and software is important for providing interactive learning in laboratories. As notable features, this book explores forces affecting the likelihood of students performing dissection in precollege classrooms. Teachers, students, and parents have points of views and attitudes regarding dissection, and all are affected by the legislative and regulatory processes and by the funding available for science instruction.

Chapter 1 begins with the crisis in science education, focusing on critical problems associated with the crisis, barriers that limit teachers in being able to do better, and the importance of science to the children and their parents in gaining the basic biology knowledge for their own scientific and health literacy and health management. To set the stage and provide an understanding of how we arrived at the current state of dissecting practice, we then offer Chapters 2 and 3, presenting historical highlights that led to current practice. Chapters 4, 5, and 6 offer an overview of the testing, guidelines, and regulatory framework affecting dissection. Chapters 7, 8, and 9 focus on the teachers, students, and animals that are directly involved in dissection. Chapters 10 and 11 concern developing and accessing new teaching resources.

Tables within the chapters provide direct access to the relevant resources with Web addresses, and each is matched with a Web-based document that can be accessed for active links to the information. Table 4.1 provides links to information on required and elective national and state science testing to meet educational standards goals. Table 5.1 provides links to national and state science and health standards and frameworks. Table 6.1a–d offers links to laws, regulations, and policies pertaining to biology instruction, budgeting for education, dissection of animals by students, and animal welfare. For teachers and others seeking general sources of information, Table 7.1 features organizations that provide information on alternative teaching resources; Table 7.2 concerns programs that loan out teaching resources for biology education. Table 10.1 focuses on specific teaching resources for biology education. Direct access to specialized databases for teaching resources appears in Table 11.1; an additional research bibliography related to dissection is provided in Table 11.2.

THE AUTHORS, OTHER COLLABORATORS, AND REVIEWERS

We, the authors, are affiliated with the UC Davis School of Veterinary Medicine. Initially a seventh grade science teacher, Lynette Hart's studies deal with human–animal interactions to better understand how humans and animals affect each other, along with the study of animal biology and behavior. Mary Wood, a biological science librarian, develops groundbreaking, Web-based searching tools in the areas of science education, animal welfare, and human–animal interactions. Benjamin Hart, a veterinarian, studies the biology and behavior of mammals.

We are indebted to David Anderson who shared his private archive of publications in the area of dissection. This book has benefited from reviews of earlier drafts of chapters by David Anderson, Jonathan Balcombe, Teri Barnato, Marc Bekoff, Laura Ducceschi, Melora Halaj, Cathy Liss, Nick Jukes, Barbara Orlans, Jo-Ann Shelton, and Bill Storm. Nancy Chan developed the two photographic montages illustrating undergraduate laboratories in comparative anatomy that are equipped with reusable materials.

Finally, we acknowledge the many teachers we have met who creatively present biology to their students. At the University of California, Davis, Michael Guinan was a consummate inspiring educator who embraced and succeeded in improving biology laboratories for undergraduate students. We have met numerous others working at the precollege level as they participated in focus groups, including Bill Storm, who brings imaginative approaches to his teaching and has helped inform us of possibilities. One of the innovative teachers we have met, we mention some of his works within the book and offer his voice as a representative of classroom teachers.

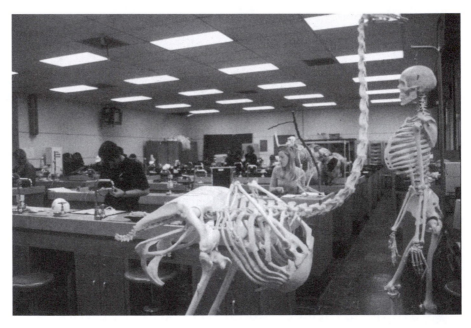

Figure P.1. Reusable fish, bird, and mammal specimens, molds, microscopic slides, sketches, and models may be used in a comparative anatomy course such as this one for undergraduates at the University of California, Davis. Different specimens are placed at 30 laboratory stations each week.

CHAPTER 1

Today's Biology Classroom Crisis

Almost all readers will recall dissecting a frog, rat, or fetal pig in junior high or high school. The question of how animals are used for dissection in precollege classrooms brings up people's memories of their own experiences in seventh grade or high school. For some of us, the experience may have been intriguing, and for others it left an unpleasant memory. The issue rings with emotional and personal perspectives, to a point that the controversy may overshadow the actual experience and benefits. Yet, despite dissection being a topic of controversy, analyses of the dynamics associated with this practice in schools are unavailable. How did dissection arise in precollege laboratories? What reasons do teachers give for planning this type of laboratory experience? What other resources are available to teachers for providing a similar type of learning? Rather, discussions of dissection generally are oriented in a debate format, as to whether it is a good practice or not. Occasionally, papers have been published as a topic of debate, in favor and against dissection, as with a supportive one written by a dean of a liberal arts college (Kline, 1995) and an opposing one written by a philosopher (Sapontzis, 1995).

This book tackles the topic of animal dissection in education, an activity that accounts for a brief period of time in the lives of most precollege students, yet one which many of them remember. Dissection from time to time becomes a subject of agitated discussion, usually as a very specific focus on this use of animals. Most of the efforts to change the practice have been spearheaded by noneducators who are acting as advocates for the animals, and have not considered the educational context in which the activity arises.

In this book, we examine the entire context, exploring the historic, legislative, and educational aspects of performing dissection in precollege classrooms, as well as considering the needs that these children will face as adults once their formal schooling is completed.

Our own experiences have been in the context of veterinary education, where we have witnessed a shift from dissection to the use of effective software learning tools. Just twenty years ago, precollege biology education was being enhanced and moved forward by the production and widespread dissemination of high quality instructional tools on biology (Hart, 1995). Nowadays, however, precollege and college biology instruction sharply differ from each other in terms of sophistication of the tools available for teaching, especially in school districts with fewer resources. We have come to believe, in writing this book, that much more is at stake than we had realized. The gap between instruction in biology and health in precollege education, at a time when adults must knowledgably and increasingly orchestrate their own health care, presents even more reasons for children to acquire information regarding biology and be given some tools to help them in their lifelong quest for health maintenance and management, and disease prevention.

This book takes a broad view, considering the various stakeholders in this issue, as well as the institutions and policies bearing on education in biology and health. Educational institutions from intermediate schools through the health sciences professional schools for many years liberally used animals for laboratories in anatomy and physiology. Many forces have sharply reduced these uses of animals in colleges and professional school courses. Veterinary schools have systematically created an array of teaching resources and clinical experiences to supplant the former terminal uses of animals in laboratories. But precollege and undergraduate institutions have not had the resources at hand to produce modified laboratories. For the most part in undergraduate education, the former laboratory courses in comparative anatomy and physiology have been phased out. While not the central topic here, this creates a crisis in that this generation of graduate and professional students knows very little about whole vertebrate organisms. Simple identification of body parts is no longer an easy task for these students who have extensive experience with cell and tissue culture, but little with whole animals.

Coming back to precollege education, just as with undergraduate education, it is hard to find an energetic infusion of resources into biology laboratory curricula that would inspire and excite students. Some of the activities that worked in the past for secondary education are no longer affordable and are no longer supplied, and new replacements have not emerged to supplant the former laboratory activities.

ANIMALS IN PRECOLLEGE EDUCATION

In this book, we seek to explore the various forces involved when teachers decide to schedule a laboratory involving dissection, and we consider other options that are near-at-hand to accomplish the same goals. We believe that most people hold the perspective that shifting to other effective teaching resources, with the same or improved learning goals, rather than dissecting animal cadavers, may be a worthwhile goal. We assert that some obstacles are delaying the shift to newer resources in precollege education, and that there is value in clarifying those obstacles. This question has importance beyond whether or not a person favors the practice of dissecting animals, as it relates to the availability of high quality biology education for our children in intermediate and secondary education. As a society, we need to be making improvements in education that foster children's biological knowledge.

Historical Convention of Dissection

The practice of dissection arose in the earliest explorations of mammalian anatomy and physiology and has continued in precollege education to the present. We set the stage for understanding the origins of dissection with humans and animals in education in Chapter 2, describing the early dissection demonstrations, held in theaters that were associated with the study and early discoveries of anatomy. These university-based practices led to the establishment of precollege biological science education, as presented in Chapter 3. Over a few centuries, a tradition and culture for the study of biology evolved, including study of actual specimens in classrooms and museum collections.

Controversy Regarding Animal Use

Strikingly, uses of animals in dissection and vivisection seem always to have been a focal point of emotional controversy. Opposition often has been spearheaded by people not involved in the educational or scientific enterprise, but rather by those advocating on behalf of animals. Many different perspectives come into play, representing all the various constituencies involved today. We feel that to move forward toward new solutions, it will be worthwhile to consider these varied perspectives. Thus, we include in this book a consideration of the perspectives of teachers in Chapter 7, the students (and implicitly their families) in Chapter 8, and the animals in Chapter 9.

Also shaping the current practices are the array of educational standards and frameworks at a national and state level that set the objectives for biological and health education, as we discuss in Chapter 5. Perhaps even more impactful today are the national requirements for educational testing—No Child Left Behind—that are the basis for significant federal funding: these are discussed in Chapter 4. Legislative, regulatory, and policy initiatives discussed in Chapter 6 influence the content of biology and health instruction, expenditures for education, and students' options to decline participating in dissection, if they prefer. Legislative constraints on uses of animals in precollege education are almost nonexistent, although optional guidelines are sometimes followed.

When controversy erupts against dissection of animals, it often is underlain by an assumption that effective alternatives are near at hand that would suffice as well in classrooms. This line of discussion may target teachers as being unwilling to use other resources (often termed alternatives), a topic that we view as more complex and related to the entire context of teachers and students. Whereas for critics, dissection may loom as a large issue, teachers may have nebulous views as to the purpose and knowledge to be gained in the dissection experience. It is likely to be viewed as an optional extra to be provided in wealthy schools, but not closely related to specific outcomes. Teachers' beliefs regarding evolution may also affect their commitment to dissection. The expectation of outcomes for dissection has not been clearly articulated among educators. Thus, teachers treat dissection in disparate ways, with some almost feverishly using it, even inappropriately with very young children, whereas others in secondary schools refuse to use any specimens or cadavers.

We describe in Chapter 10 the very large number of alternative resources for teaching, and offer suggestions in Chapter 11 for efficiently searching to locate desired resources. Better use of these resources is one path forward, and we offer several tables aimed at bringing the resources more easily within range.

Eight tables complement the chapters, complementing the topics of testing, standards, and legislation, providing information for teachers on helpful organizations and loan opportunities for resources, listing sites for teaching resources and Web sites, and listing additional useful information sources. A companion Web site provides similar information online, with links to the indicated resources, http://www.vetmed.ucdavis.edu/Animal_Alternatives/appendices.html.

EDUCATIONAL CRISES

Almost daily, newspaper headlines emphasize problems in primary and secondary education, especially science. Systematic studies and evaluations have made three points: (1) U.S. students are slipping behind in performance (Trends in International Mathematics and Science Study, 2004), leading to national efforts to raise standards and achievement. (2) Young people from lower economic settings are not attracted to higher education (U.S. Department of Education, 2002a). (3) The quality of current laboratory experiences seems to be poor, with assessment of student learning needed (U.S. Department of Education, 2002a). Secondary school science laboratories are under fire for lacking clear goals, not engaging students in discussions, and failing to illustrate how science methodology leads to knowledge (Singer, Hilton, and Schweingruber, 2005). These discouraging trends are especially critical because for many students, high school offers the last organized instruction they will have concerning their own physiology and health.

Problem 1. U.S. Slippage in Science

We need look no further than *Time* magazine with the cover featuring, "Is America Flunking Science?" and a lead story entitled, "Are we losing our edge?" (Lemonick, 2006). As evidence of the problem this analysis pointed to shrinking U.S. government spending on research, a tapering off of articles published in science and engineering journals, fewer doctorates in science and engineering fields, and a low percentage of twenty-four-year-olds with science degrees.

International testing of children in grades four and eight provides an opportunity to compare the learning of children in various countries. In general, Asian countries have the highest scores, Singapore, Chinese Taipei, Korea, and Hong Kong leading in grade eight in the 2003 test, and Singapore, Chinese Taipei, Japan, and Hong Kong leading in grade four (Trends in International Mathematics and Science Study, 2004). The U.S. eighth graders scored ninth in science in the 2003 test, improving from seventeenth place in 1995, and the fourth graders slipped from third to sixth place.

National test results in science testing recently showed some improvement at grade four, but no change at grade eight and declines at grade twelve, as analyzed in "The Nation's Report Card: Science 2005"; minority students in grades four and eight showed improvements (National Center for Educational Statistics, 2006).

Problem 2. Socioeconomic Slippage in Higher Education

In spite of significant efforts to recruit and financially support students from lower income and racially diverse backgrounds, figures reveal that we are not making gains in keeping a broad range of students interested in entering higher education in the sciences. The slippage is already well established in the precollege years, where lower-socioeconomic students perform less well, and are less academically prepared to enter college. The deficits go beyond just academic preparation, as the lower socioeconomic students, as well as the African American and Latino students, also are less physically fit, and show other liabilities in fitness that are likely to compromise their performance.

Problem 3. Unfocused and Ineffective Science Laboratories

While much of the emphasis in the public arena and the educational professional literature in recent years has been on diversity and supplying equal access to education, attention has shifted away from laboratories, to such an extent that state standards and testing place little emphasis on laboratory competencies. Integrating laboratories to be consistent with the science standards and frameworks has not been a priority, nor providing the resources required to offer effective laboratories.

A recent comprehensive analysis of high school science laboratories defined their purposes, which include: enhancing the mastery of subject matter, developing scientific reasoning, understanding complexity and ambiguity of empirical work, developing practical skills, understanding the nature of science, cultivating interest in science, and developing teamwork (Singer, Hilton, and Schweingruber, 2005). The report flatly concluded that the quality of current laboratory experiences is poor for most students, and went on to suggest major changes in undergraduate science education to improve science teachers' capacity to lead laboratory experiences, as well as providing more comprehensive systems of support for teachers. However, sometimes high school administrative structures, science standards, and current testing work against improving laboratory experiences. Assessment is needed specifically of effective teaching and student learning in laboratories. Learning goals could better be achieved if laboratory experiences were designed with specified learning outcomes, were sequenced into science instruction with an integration of science content and processes, and included student reflection and discussion.

As the report documents, the primacy of testing as the measure of success has resulted in less administrative support and funding for offering laboratories, and less clarity about what might be effective methods for teaching in laboratories. This imbalance hampers the capability of teachers to deliver laboratories that are well integrated with the other course material.

TEACHERS

While on one hand particular pressures and criticism are directed at teachers for the poor performance of students on examinations, they may be criticized by others for providing dissection in classrooms. Efforts in teaching biology to students today present distinct challenges for the teachers, as well as the parents and children who are involved. Teachers are sometimes blamed for the limited use of newer resources in secondary education, which so sharply contrasts with the widespread creation of new teaching tools in professional medical education (Hart, Wood, and Weng, 2005). But, examining this problem from a teacher's perspective, we find that three barriers obstruct mainstreaming of new teaching resources in high school biology courses to replace or complement dissection (Hart, Wood, and Weng, 2006). First, dissection is not addressed in major educational documents that include course outlines, curricular standards, and frameworks. Secondly, resources that formerly were provided for science education in school districts, supplemented with financial and technical support, are no longer available. Thirdly, teachers actively search for motivating materials that excite their students about learning; but such materials are difficult to find today, and the classroom time frame available is very limited.

Barrier 1. Dissection Not Addressed

Although teachers have various science frameworks and curricular standards as guidelines, the practice of dissection is seldom mentioned. These resources focus on the testable vocabulary and definitive concept language to be taught, without addressing the specific laboratory lessons to be used. One might expect that a somewhat controversial topic such as dissection, which has even been the focus of some legislation, would be a topic for lively discussion in educational materials and philosophy of education texts directed to precollege. On the contrary, the topic is not prominent—and is almost difficult to find in the literature. Major dialogs concerning science curricula do not consider dissection, nor does the topic arise in course outlines. We were unable to find an ongoing prominent platform where teachers and educational professors are discussing teaching methods for biology laboratories, and whether they involve dissection or alternatives. Teachers are left to solve these problems for themselves and figure out a course of action in developing their laboratory exercises.

Barrier 2. Resource Centers Phased Out

New classroom teachers preparing to teach science used to be surrounded by supportive resources provided by the school district. In past decades, county or city school districts maintained audiovisual teaching aids centers with resources required for science laboratories. Teachers could easily order these materials in advance for their classes. Curricular libraries included an array of textbooks and related supplementary materials. Each recommended course of study was laid out in teaching manuals with detailed lesson plans that were coordinated with the required laboratory materials that could easily be ordered. Experienced mentor teachers were available to assist newer teachers in figuring out how to organize their laboratories and acquire the materials necessary for each scheduled laboratory. Budgets were designated for teachers to purchase essential supplies. Over the years, the budgets for these supportive resources have been sharply reduced and the resource centers that provided laboratory materials ultimately were dismantled.

With the decrease of funding and a shift of resources to testing, teachers are expected to assume the responsibility of figuring out how to assemble laboratory exercises with the necessary materials. Being somewhat forced to hoard materials reduces collegiality. Increased teaching assignments across the board result in teachers having less cooperative preparatory time for teaching. As they teach over a period of years in a climate of increased isolation, teachers acquire some reusable resources, plus they may use their own funds to purchase some materials to make the laboratories more interesting. Heightened costs of materials involved in dissection laboratories these days make it unlikely that teachers will try to self-fund their biology laboratories. Although commercial products are available, they are costly and not presented as an integrated set of resources for high school biology (Weng, Wood, and Hart, 2004). Teachers can borrow some resources from Animalearn (2007) or the Humane Society of the United States (2007a), and other sources as discussed in Chapter 7, but this requires farsighted planning for particular lesson plans.

Barrier 3. Teachers' Efforts to Motivate Students

College students enter the field of education because they enjoy the contact with children and the rewards of being part of children's development and seeing them learn. Teachers live for the excitement of inspiring students to learn. They are generous with their time and money in working to enhance learning of their students, and become disheartened when they cannot help their students make progress. Their job satisfaction is eroded when the learning context is compromised. Too many discipline problems, lack of parental support, and an absence of essential teaching materials can degrade the learning environment from the teachers' perspective.

In the realm of science education, teachers should reasonably expect to provide laboratories that enhance the subject material and get students excited about learning science. But it is often on their shoulders to acquire materials to provide a good laboratory experience. Laboratories coordinated with text material, and equipped with necessary supplies, are not generally at hand, nor are they easily produced. Even for those few teachers who have a reasonable budget to order materials, the available laboratory resources are not well coordinated with the teachers' instructional needs and schedules.

Teachers like using effective materials that excite and interest students, and as mentioned, many use their own funds to purchase materials to enhance their teaching. A bit of creative financial investment such as this by teachers goes far for someone teaching in the lower grades. However, for instruction in biology with studies of the physiological systems of mammals, a few dollars cannot supply effective laboratories and teaching models. Over the years, science teachers gather up useful materials that they acquire as they go along, some contributed by parents, some purchased by the teachers, and some acquired from other teachers. Inexpensive, high quality, and widely accessible teaching materials would be a major addition to help support science and biology teachers in their quest for effective teaching. The teachers' goal of inspiring students in biology can seem unattainable, even though so important. The teachers know that biology is an essential area of knowledge that relates to students' own bodies and medical care, and that, for many students, their precollege courses will be their only instruction pertaining to their own biology.

Some further challenges beset science education. In the lower grades, most teachers have not been science majors. Many feel poorly prepared to teach science lessons. Most of those teaching in the intermediate and secondary grades have academic preparation in the sciences, but it is common to hear of them being intimidated by new teaching materials or technologies, or disheartened by the dearth of creative, interesting content. They arrive in the classroom with little support available from the district, and hope for help from opportunities in workshops or continuing education. Whereas older teachers may be struggling with the new electronic world, younger ones accustomed to the high-speed learning environment surrounding everyone may be shocked at the confined presentation space of the classroom. Biology is an exciting topic—the living world that surrounds children. It is a natural topic of curiosity, engagement, and sometimes disgust. All of these can be harnessed for motivating children in the classroom.

Significant constraints are imposed by the daily teaching schedule. On the surface it may appear to have eased up for teachers in recent years with hiring of yard and recess duty personnel. Yard and recess duty used to be a typical part of the teaching day, actually providing an opportunity to see the children in a different, more relaxed context. While some assistance is available, still teachers are expected to perform a variety of adjunct duties that may include performing as crossing guard or monitoring on yard duty at recess. The numbers of classes per day, and numbers of students per class, vary today as in the past. A dramatic increase in standardized testing impacts all classrooms, not only on those days of tests, but also throughout the year, as it seems imperative for test scores of classes to continue to improve, creating an inevitable requirement that teachers to some extent teach for the tests. Usually unmentioned is the high cost of the testing enterprise, which drains funds that could otherwise be used for enhanced teaching or laboratories.

CHILDREN

For the students, school days have become more jam-packed and hectic, often with a school bus or car-pool commute. Their schedule may not provide them with physical education and exercise, nor do they typically walk or bike-ride to school as in the past. Watching television in the evening may be a major pastime, rather than self-directed activities such as reading or playing outdoors in neighborhoods. Although computer-savvy and accustomed to fast-moving visuals, many children make their way through high school and into fine universities without mastering basic math and writing skills that are a part of the scientific process. Entertaining diversions such as television or video games may supplant the learning skills, study habits, and initiative that children otherwise would have developed as they pursued their individual interests. Fine motor skills formerly honed by sewing for girls, and model building and shop for boys, are undeveloped in many children today.

Biology has a potential to be one of the most interesting subjects for children—learning about their own bodies. In a world where crimes against children are so visible and sensational, the societal focus emphasizes perversity and abandons and even hampers basic instruction that is essential for health awareness and maintenance. In addition, the instructional and laboratory materials the children are offered in classrooms fall short of what could be offered with newer instructional technology. The virtual software and imaging methods today create almost limitless possibilities that already have been harnessed for Hollywood productions and videogames. Even a portion of such technology, as already available in medical and veterinary education (Hart, Wood, and Weng, 2005), would update the learning environment and provide valuable learning opportunities to children concerning their own bodies and health.

Equal Access to Learning Biology and Health

Unlike basic reading and writing, learning in the sciences is facilitated with laboratory and practical experiences that provide exposure to the technology and instrumentation, as well as the methods of data collection, analyses, and presentation. Such laboratories are effortful and expensive to construct and equip, and to supply with the necessary expendable materials. Children's access to effective learning environments for science varies with the number of the children in the classroom, the preparation of the teacher, the facilities and resources available for teaching, and the appropriateness of the level and presentation of the material. Schools in lower social or economic neighborhoods (i.e., low tax base), or that are overwhelmed with behavioral or linguistic challenges in the classroom, are less able to deliver an outstanding presentation of biology and health and facilitate children in learning this material.

Last Chance for Basic Preparation

Most children continue their formal science education through at least the seventh grade general science course, and many complete a high school biology course. Some aspects of health are covered in a separate course, sometimes coordinated with driver training or physical education. A significant proportion of children do not complete high school; about 71 percent graduated nationally in 1998 (Greene, 2002). A wide disparity was reported among racial groups, with a rate of 78 percent for white students, 56 percent for African-American students, and 54 percent for Latino students. Geographically, the rates ranged by state from 54 percent in Georgia to

93 percent in Iowa. African-American students in Wisconsin had a graduation rate of 40 percent, while those in West Virginia had a rate of 71 percent. Among Latinos, those in Georgia had the lowest rate, 32 percent, while those in Montana had a rate of 82 percent. Among the fifty largest school districts in the United States, the overall graduation rate ranged from 28 percent in Cleveland, Ohio, to 87 percent in Fairfax County, Virginia. Thus, in some geographic areas, a majority of students, or some subgroups of students, do not continue formal education after high school.

Among high school graduates, in general fewer than half of graduates complete the requirements of college preparatory courses. As an example, in San Mateo County, California, only 43 percent of public high school graduates complete the courses required for University of California or California State University admission, with a grade of "C" or better (Lucile Packard Foundation for Children's Health, 2006). Thus, despite the increased proportion of children entering college in recent decades, the intermediate and high school years are the last opportunity to reach virtually all students and offer them basic information on biology and health.

PARENTS

As in the past, parents value and place a high priority on their children's effective school performance and are motivated to see their children excel in school. But, family profiles have changed, with more single-parent households and parents with dual careers. Less energy and relaxed time are available for involvement with children and their classwork. Fifty years ago more parents and neighbors were at home to greet their children at the end of the school day, and could monitor daily school progress more closely. Many children walked to their nearby neighborhood schools. Their parents knew the teachers and often kept close tabs on their kids' school performance. Parents and neighbors could monitor the neighborhood after school activities. When problems arose in the classroom, it was typical for parents to support the teacher and follow through to assure that the behavior and school performance of the child would improve. Today's parents do not necessarily support the teachers' reports. Parents may not insist their children improve their behavior and achievement. Teachers cannot assume that they can develop an effective partnership with parents so as to improve a child's performance.

Some small proportion of parents still provide close oversight of their children's classroom performance. Many of these parents are proactive in obtaining outstanding teaching resources and providing them at home—an array of materials that may exceed what is offered in the classroom. In upper middle class and well-funded college towns, it is easier to find parents who have worked with their children at home using high-end anatomy software such as A.D.A.M. (2007a, b), than it is to find teachers who have had access to such software in the classroom. One hears of a wide disparity in the at-home instructional support that parents provide. The disparity inevitably arises from the range of parental education and knowledge, and is furthered by the resources of available parental time and money.

A growing number of parents in recent years have been opting to home school their children, even through the high school years. Home schooling has its own challenges in providing biology laboratory instruction, despite the close oversight provided by parents. Materials designed for this specific purpose are available for purchase. Parents who have the means can order materials that may, in fact, provide the equivalent to what is offered in their public school setting.

Role of Families in Informal Science Education

Probably no one is as motivated to teach a particular child as the child's parent. Families are invested in helping their children succeed in school and move ahead. Family traditions for reading and learning set a pattern for children to learn to read, enjoy reading, and love learning (Taylor, 1998). A large proportion of families make the effort of taking their children to zoos. While there, they adopt a conversational instructional role in the zoo with content that often is indistinguishable from that of teachers (Tunnicliffe, 1996c). These ethnographic, observational studies demonstrate the important role of parents and other family members in setting a pattern of learning and supporting it throughout development. Families may get involved in helping with homework, assisting in field trips, helping with at-home projects, and coaching the children in study skills.

With the availability of television and computers, families are involved in making some choices regarding their children's access to media. They may employ readily available computer programs or software, so that opportunities for at-home learning, fostered by parents, are richer than ever before.

Experiences of Families with Health Management Decisions

Any extended family provides exposure and experience with some family members who have sickness and serious disease that require them seeking medical assistance and making decisions. These issues are so compelling today that health has become the top domestic issue on the minds of Americans, according to a survey by Research!America (2007). The hereditary component of many diseases increases children's interest and motivation to learn and adds personal value to them for gaining knowledge about health management and disease prevention. Common diseases such as diabetes, heart disease, and cancer become more compelling topics if a family member, even a pet, is grappling with them. The personal link can be an added component toward motivating the student to learn more, at the very time as the student is observing the family sorting out decisions on medical care and treatment with physicians.

CURRENT SITUATION

In contrast to the small pool of early scholars who studied animals long ago, today's body of knowledge is widely disseminated, potentially to everyone. Nowhere is this more dramatic than in the exhibitions of human plastinated bodies by von Hagens (BODY WORLDS, 2007a), that have appeared in major cities in Europe and the United States. Artistically presented as performing athletes, the skeletons, muscles, and organs of human bodies are splayed out for best display, with the skin held separately, or as a series of slices offering a see-through view that cuts across all the organs (von Hagens and Whalley, 2002). These BODY WORLDS exhibits are presented in vast galleries, interspersed with attractive potted plants, drawing huge crowds that closely examine the details of each presented body part. A few bodies of animals are also on view. In BODY WORLDS, von Hagens displays a wall of controversial newspaper articles attacking his own exhibit, meanwhile arguing that his work effectively democratizes knowledge about the human body. The exhibits are enriched by a comprehensive catalog and many pictorial games, videos, and other resources. What had been secret and mysterious except for rare and occasional access is now open for scrutiny, eliciting fascination from some young children as well as adults who are also fascinated with biology and dissection. The exhibition even includes an opportunity to viewers for written commentary and reaction to the exhibit. This educational tool clearly reveals the grey lungs of smokers as they contrast with the lungs of nonsmokers, and also presents diverse examples of cancers, and joint replacements—offering an internal view.

The supply of teaching resources is ever growing and discussed in Chapter 11, and these are fully searchable in the NORINA database (Smith, 2007). InterNICHE (International Network for Humane Education: Jukes and Chiuia, 2003; InterNICHE, 2007) offers a book in many languages and a Web site categorizing some alternatives. The searchable AVAR Alternatives in Education database provides access to additional information (Association of Veterinarians for Animal Rights, 2007). Perhaps the most obvious sources of information are those offered by commercial distributors, who advertise extensively and send out catalogs, keeping teachers well informed regarding their products (Carolina Biological Supply Company, 2007; NASCO Online Catalogs, 2007). Despite all these resources, teachers do not have ready access to practical materials that interface well with the subject matter and offer the excitement and flexibility that would characterize superb resources for precollege biology and physiology.

Considering the rapid changes in uses of animals in higher education, it seems surprising that production of new resources supplanting dissection of animals in precollege education has been so slow-moving. The technology is available to create extraordinary educational materials on software and make them widely available to reach children in every community. Veterinary schools have been engaged over the past couple decades in a process of increasingly mainstreaming resources such as software and models that do not require a consumptive use of animals (Hart, Wood, and Weng, 2005). By now, dissection is largely phased out in veterinary education. Paradoxically, though, dissection persists in precollege education.

Although educators in recent years have focused their research and writing primarily on other topics than dissection, strong opposition to the dissection of animals in precollege classrooms has arisen from animal advocates. Reviews have addressed the ethical considerations relating to uses of animals (Langley, 1991), and the patterns of use of animals in the United States (Orlans, 1991), and in higher education (Balcombe, 2000). Some scientists have provided thoughtful consideration of the adverse ethical consequences that may arise from instructing students to work with animals that have been killed for their education (Orlans, 2000). Generally, these analyses have emphasized political or philosophical perspectives, often presented in debating or argumentative formats, rather than educational ones. Major efforts have focused on supporting students who prefer not to participate in dissection, offering them counseling or legal support (Balcombe, 1997), and sponsoring legislative efforts to give them protection. A major difficulty with this line of effort is that teachers have not been engaged as leaders informing these discussions. The published discussions thus have not benefited enough from consideration of the teachers' perspectives and challenges in teaching.

One important question, especially in the early days of teaching innovations, was whether the innovations are educationally effective when compared with dissection. These studies also revealed students' attitudes concerning their experiences in dissection, which ranged from positive and exciting, to negative and leading students away from biology. Studies from England surveying students concerning their experiences and attitudes toward using animals in education (Lock and Millett, 1992), and also teachers concerning their experiences and attitudes (Adkins and Lock, 1994), revealed that about one third of the teachers' opinions were leading them away from using animals in teaching. A study of thirty-four Australian schools revealed that all offered dissection, plus usually activities with live animals; cost was the major limitation inhibiting offering dissection (Smith, 1994). A study of Canadian students' earlier experiences and attitudes concerning dissections performed in secondary school revealed that a substantial minority retained primarily negative attitudes concerning the experience (Bowd, 1993). While much in agreement, Lock (1994) differed in feeling that no alternatives were superior to dissection. A study of the reactions of students to their experiences in dissection found that a significant minority had negative memories of the experience (Barr and Herzog, 2000).

Both societal and classroom changes over the past fifty years have impacted the current science or biology classroom in grades seven to twelve. Paradoxically, one can argue that despite the abundance of media capabilities, the learning environment and resources available for teaching science have shrunk as compared with the days of "ditto" machines. How can this be? Where is the social will to sustain and nurture the minds of our young children? Why do we not consider seriously increasing resources devoted to teaching science so as to enhance children's love and capacity for learning?

Physical fitness at school, junk food in school cafeterias, and basic skills in arithmetic are among the controversial topics in American schools today. Decades ago, Americans had a strong reaction to the Russian achievement of Sputnik, and America focused its will for educating children in science. But today, fewer students are entering the sciences, and America is at risk of losing its excellence in the sciences. In a biologically knowledgeable society, people are more prepared to make wise choices for nutrition, health management, and lifestyle factors. The costs of shortcutting biological science education are especially far-reaching. More than ever, people are finding it necessary to be vigilant in managing their own access to medical care and making medical decisions. For children who only complete high school, a basic education in biology is their only chance for a foundation in being knowledgeable consumers of medical care, while practicing health maintenance and disease prevention.

It seems somewhat surprising, that dissection persists today in precollege classrooms despite its seemingly questionable educational value and being out of step with ongoing trends in university education. We address in the coming chapters the variety of forces that have established, and continue to sustain, the common practice of dissection in classrooms.

USING THIS BOOK

We expect that most readers will be interested in selected portions of this book as a specialized reference on certain aspects related to dissection. With that in mind, we have made some effort to prepare each chapter with stand-alone text, but we still mention other related chapters in the book as appropriate. Thus, for example, references to loan programs for teaching resources appear in several chapters.

Providing convenient access to relevant information is a central goal of the book. Thus, as a unique feature of this book, we include a variety of tables on many aspects of dissection, and a new Web site that includes some stored searches relevant to dissection. Although the citations are listed in the references, the array of eight tables provides streamlined access to key resources on the major topics of the book, including testing, educational standards, legislation, relevant organizations, teaching resource loan programs, teaching resources, Web sites, and key resources. As

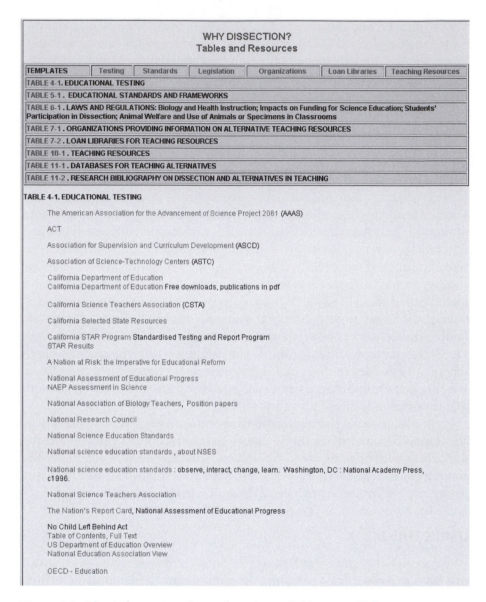

Figure 1.1. The information shown here is available on a Web site companion to this book, available at http://www.vetmed.ucdavis.edu/Animal_Alternatives/ appendices.html. Titles for the 8 tables are listed at the top of this web page and each title is linked to that specific table. Table 4.1, as partially shown here, includes links to resources pertaining to educational testing.

described in Chapter 11 and shown in Figure 1.1, a complementary Web site mirrors the eight tables for those wishing to make use of active Web links and searching tools to some resources, databases, search templates, and searching tutorials, http://www.vetmed.ucdavis.edu/Animal_ Alternatives/appendices.html.

CHAPTER 2

Early History of Dissection: Controversy and Advances

How did we come to our current position where dissection is practiced in high school biology classrooms while being phased out in professional medical and veterinary schools? Examining how anatomical knowledge was originally acquired from dissection and taught to students of medicine explains, at least partially, why dissection continues into the present in formal education at the K-12 level. There is perhaps a disconnect between advances in the study of anatomy at the professional school level and at the precollege level where the newer alternatives to dissection have not filtered down. It is critical to remember that the practice of dissecting animals is connected with dissecting human bodies. In this regard, perhaps the main theme of the history of dissection is contentiousness, much of it stemming from emotional reactions and religious considerations. Chief among the emotional resistance to human dissection was the horror at having your body cut up after death. For Christians, being cut up and deprived of a proper burial set the stage for problems later because the person is expected to be resurrected, a main tenet of Christian belief.

Interestingly, despite these emotional overtones, the practice of dissection in the Middle Ages and during the Renaissance often involved cooperative agreements with church leaders. Enlightened religious leaders seemed to have recognized the value to human welfare of the scientific information acquired from dissections, at least if conducted with respect for the deceased. Nonetheless, dissection of human cadavers requires suppressing a normal emotional reluctance to engage in the "mutilation" of another person's body (Richardson, 2000). For some, this requires denying ordinary emotions (William Hunter, in Richardson, 2000).

While the primary interest was in human anatomy and the benefits to people, animals were frequently used as substitutes or surrogates for the sake of learning basic human anatomy. By cutting into a dead human or animal, one assumes that the functioning of the whole body of a live person can be understood by examining its parts in a dead individual (Cunningham, 1997). Furthermore, when dissecting animals, there is an expectation that the observations of the animal's body parts will apply to human counterparts.

The focus of this book is the dissection of cadavers. It is appropriate also to mention there has always been a debate surrounding the study of live animals where parts of the body are examined while the animal is still living. The practice of cutting a living animal to show the functions of the body is referred to as vivisection. In his classical writings in the second century after Christ, scientific pioneer Galen of Pergamum, and physician to the emperor Marcus Aurelius, emphasized the advantage of studying animals while they were still breathing and argued that more accurate information about the function of various organs could come from such vivisections. Much later, sixteenth century pioneers of medicine, including Andreas Vesalius and Realdo Colombo (also, Columbo), echoed Galen's viewpoints (Maehle, 1993). Whether vivisections were ever performed on people in this period is not well-documented, but such practices were viewed as a crime and were undoubtedly very rare. While animal vivisection continued to be viewed by many as an alternative to understanding human bodily functions by the eighteenth century, concerns were raised by Alexander Adam von Sinclair and Samuel Johnson that vivisection of animals could lead medical students away from compassion and towards cruelty, hardening their hearts to human life (Maehle, 1993).

DISSECTION AS A SOURCE OF KNOWLEDGE: FROM ARISTOTLE TO THE ITALIAN UNIVERSITIES

A genius with a prodigious output in virtually every field of knowledge, Aristotle of Macedonia pursued his studies three centuries before Christ and was a pupil of the philosophical giant, Plato of Athens. Aristotle was hired as a tutor for Alexander the Great when he was still a child. Aristotle's findings in biology drew from his dissections and the techniques along with anatomical information are presented in his book, *De Anima*. Advances in biological information following Aristotle were extremely slow. His work stood unchallenged for hundreds of years. Long after Aristotle provided a wealth of new information, it was Galen, working in the second century after Christ, who compiled available anatomical knowledge in a medical treatise that was then disseminated to medical schools in Alexandria, Byzantium, and to the western Roman Empire. This work synthesizing the work of earlier scientists basically concurred with Aristotle's findings.

It was not until about 1000 years after Galen that progress was made in the study and teaching of anatomy as learned from dissection. The practice of dissection is documented by the Italian physicians at the medical school at Salerno in the thirteenth century (O'Malley, 1964). In the fourteenth century at Bologna, anatomist Mondino de' Liuzzi wrote a text on anatomy that was used in teaching, based on his own dissection of human bodies as well as material from Galen's writings. As a physician, he taught anatomy to medical students by human dissection. His philosophy was that rational learning and special knowledge of medicine was the province of medical doctors. An interesting twist in the teaching of anatomy was that the professor called on technicians, so-called barber-surgeons, who took his instructions for cutting into the cadaver. Intriguingly, although surgery was conventionally practiced, teaching of surgery was not highly regarded in the medical field (French, 2000). Hence, surgeons participated as dissector technicians. Only the professor had the specialized knowledge and status to direct the dissections. The slow pace with which basic medical anatomy progressed reflected the perspective of the teaching of "classical" material with profound and overwhelming respect for the early anatomists, particularly Aristotle. Scientific scholars of that day referred to themselves as pygmies standing on the shoulders of Aristotle. Perhaps indicating the perspective that anatomy does not change over the years, the viewpoint was that the existing work was preeminent and complete for all time.

Human dissection in academic settings became very popular to student spectators; nonetheless, only twenty students were allowed at the dissection of a man and no more than thirty at that of a woman. A pioneer of teaching by dissection was Gentile da Foligna who introduced this aspect of teaching at the University of Padua in the fourteenth century, with subsequent human dissection used in teaching at Italian universities in Venice, Florence, Siena, Perugia, Genoa, Ferrara, and Pisa.

While dissection of humans aroused the most interest and led to many discoveries regarding human anatomy and physiology, dissection of animals was also informative and less fraught with contention. Cadavers of animals were regarded as being in an entirely separate category from those of humans, as reflected in Descartes' (1637) writings in the seventeenth century, describing animals as being unable to think or feel pain, and, in fact, as similar to machines.

Acquiring sufficient human specimens for these demonstrations was always a particular challenge. This led to attempts to acquire cadavers by grave robbing. The practice of grave robbing became more lucrative and was in full force by the fourteenth century. When four medical students were prosecuted for grave robbing, seeking to acquire specimens for dissection, it was evident that cadavers were in short supply.

By the fifteenth century, at the dawning of the Renaissance, access to human dissections was made available to nonmedical students and introduced at Bologna and Padua, using cadavers provided by judicial authorities (Carlino, 1999). These dissections were performed by barber-surgeons, presumably to illustrate the material in the books read by the students. Like other dissections, these dissections did not acknowledge the possibility of acquiring new anatomical knowledge, but rather the dissections were carried out according to a formalized ritual of a public lesson to illustrate the writings of earlier scientists.

Drawings in wood cuts, such those in Mondino's text, show a dissector dressed as a barber-surgeon who holds the knife and performs the dissection (see Figure 2.1). The drawings show a *demonstrator* who holds a pointer and indicates to the dissector where and how to cut based on what is being said from the pulpit. Standing in the pulpit is the professor, referred to as the *lector* who, without approaching the body, recited passages from the classical anatomical texts, by Galen or Mondino.

Around this time, barbers and surgeons in England were united by Royal Charter and granted by Henry VIII, the right to the bodies of hanged felons, giving them a type of legal monopoly to anatomical material (Richardson, 2000). Given the somewhat less-respected manual work of dissection and the study of anatomy in general, academic teachers of anatomy were not yet established and the subject of anatomy remained of secondary importance (Carlino, 1999). In fact, lower salaries were paid to dissectors and, in addition, there was a requirement that they also teach another subject of a more established status.

By the time the Renaissance in Europe was in full bloom, the study of anatomy and physiology rose to a new level along with the arts and almost all areas of human endeavor. It was the desires of artists who wished to study human anatomy to improve the accuracy of their portrayal of figures that added further impetus to the study of anatomy. For example, Leonardo da Vinci, who became justifiably famous as both an artist and a scientist in the fifteenth century, worked in a mortuary in Florence and reportedly dissected some thirty human corpses (French, 1999). He was an accurate painter of animals as well, and he studied the bodies of animals after they were killed at the slaughterhouse (Belt, 1955).

An Italian artist of equal fame was Michelangelo di Lodovico Buonarroti Simoni who also worked in the fifteenth century. He was known to dissect animals and humans extensively. Although Michelangelo considered writing a book on anatomy, he was said to have given up dissecting human corpses because it so affected his stomach that he could neither eat nor drink (Cunningham, 1997). An interesting milestone in the history of medicine and anatomy was the insistence of artists to be able to accurately portray bodies that led to the motivation of medical educators and students to continue to advance knowledge by human dissection.

DISSECTION AND SCIENTIFIC ADVANCES

A turning point in the history of dissection and anatomy began in the sixteenth century with Andreas Vesalius, the Flemish anatomist, who elected to dissect the cadavers himself rather than direct a barber-surgeon while reading from an ancient text as was done by other anatomists (Carlino,

Figure 2.1. This dissection scene from 1316 shows Mondino dei Luzzi, and is typical for the period of early medical advances in showing him as a professor on a throne, directing the course of the dissection from a distance, while reading from a classic text. Working quickly before the decay progressed in the unpreserved tissues, especially in the abdominal organs and contents, was important. The initial dissecting cut here is made into the abdomen, where putrefaction was most rapid. (Photo: National Library of Medicine)

1999). Dissection became an opportunity for exploration and learning, rather than simply a demonstration of ancient knowledge. With his new orientation and his classical book, *De Humani Corporis Fabrica* (*On the Fabric (Construction) of the Human Body*: Vesalius, 1543/1964), Vesalius' work supplanted the long-established primacy of Galen in anatomy. The value of his hands-on involvement yielded important results. Not only did he correct many of Galen's errors that seemed to be based on dissections of animals, but he also discovered the distribution of the nerves of the circulatory system and the relationship of the heart beat to the pulse.

Following up on his discoveries concerning the circulatory system, Vesalius sought to teach students what he had learned by lecturing while performing dissections (Klestinec, 2004). He set forth a research method which affirmed the educational potential of dissection as a method for verifying anatomical knowledge (Carlino, 1999). For Vesalius, research and teaching were linked, as he instructed students by dissection even as he continued to learn more about the body's function while dissecting (Figure 2.2a). Through his work, and that of contemporaries, the illustrations depicting human and animal organs and bodies became refined and more accurate. Vesalius somewhat made light of the dissection scene in his Putti illustrations (Figure 2.2b). A surprising aspect of the accounts of these hands-on dissections and demonstrations is the unruliness and boisterous language of the students. Even though they were interested in the dissections, their behavior sometimes took on a theatrical or carnival atmosphere.

Prior to Vesalius, anatomy was mostly thought of as a theoretical branch of knowledge that reflected both medicine and philosophy (Carlino, 1999). The ancient scholars were viewed with religious respect, and their writings treated almost as scripture. The dissections that were done were simply treated as demonstrations for the narrative of the ancient writings that had been consolidated by Galen. Vesalius argued, in his new approach, that the cadaver provided the evidence of anatomy and that the dissection itself, rather than the writings of the ancients, was the accurate source of information. Vesalius asserted these points as a young man in his twenties, stepping well above his designated role of performing the dissection while the professor was reading the classical anatomy texts. A famous anatomy lecture by Vesalius in Bologna, described in a detailed account by Baldasar Heseler, portrayed Vesalius' daring criticism of Galen's writings (Carlino, 1999). As a young man he audaciously attacked the classical pattern of teaching and asserted his position and methods. He boasted, in *De fabrica*, of his methods of obtaining bodies. His boasts led to the enactment of laws against grave robbery.

Based on the dissections he conducted, Vesalius prepared extensive drawings, providing a repertoire of images from which people could learn about their anatomy. He positioned his new anatomy drawings as bodies in picturesque, realistic settings and poses, such as a muscled man with a city or park in the background (Figure 2.3a–d). This style of presentation provoked

Figure 2.2a. The frontispiece of the 1543 first edition of *De Humani Corporis Fabrica Libri Septem*, by Vesalius. The artist, perhaps Titian's pupil, Jan Steven van Calcar (Magruder, 2007), shows Vesalius conducting the dissection himself and facing us as he exposes the uterus of a female cadaver to a throng in the temporary, wooden anatomy theater (Dickson, 2001; Bambach, 2002). The human skeleton, centrally placed, is a reminder of human mortality. A dog and a goat in the lower right corner perhaps allude to dissection of animals. (Photo: National Library of Medicine)

Figure 2.2b. Vesalius revealed details about the process of dissection in the elaborate historiated initials of his book. In this letter Q, a live pig is shackled during a
vivisection experiment (Magruder, 2007). The procedure is conducted by cherubic
putti, little cherubs who are conducting the tasks of an anatomist. (Photo: National
Library of Medicine)

students of the day to learn more. A realistic presentation of anatomy is
being revisited even today by BODY WORLDS with the presentation of
real human plastinated cadavers presented in athletic poses (Von Hagens
and Whalley, 2002). The new culture launched by Vesalius maintained that
students could learn from their own experience in dissection by witnessing
the cadaver before them, for example a head with brain, rather than relying
on the classic textbook drawings (Figure 2.4).

   The successors to Vesalius continued performing the anatomical demonstrations of dissection as a teaching method, while making further discoveries and gradually building an array of more accurate pictorial images.
The pulmonary circulation was discovered independently in the sixteenth
century by two anatomists, Realdo Colombo (also known as Columbo)

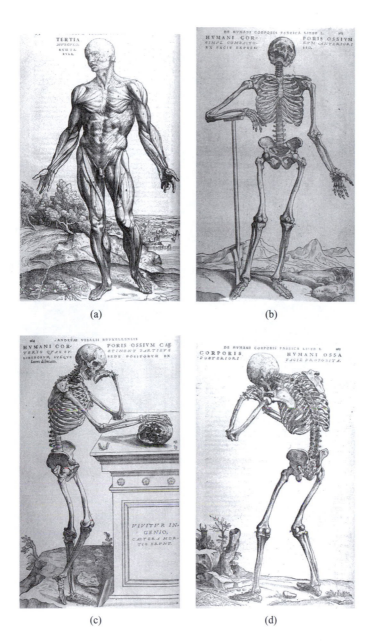

(a)

(b)

(c)

(d)

Figure 2.3a–d. The poses presented in *De Humani Corporis Fabrica* by Vesalius (1543/1964) blend anatomy and the profound questions about human mortality. The vibrant muscular figure in the top left illustration captures the joy and strength of human life. The other three skeletal figures convey the sadness and grief that may come with old age, and serious thoughts about the meaning of life, thus brilliantly linking dissection to the deepest human emotions. Similar postures and settings are used in current displays of plastinated human bodies as presented in BODY WORLDS (2007a). (Photos: National Library of Medicine)

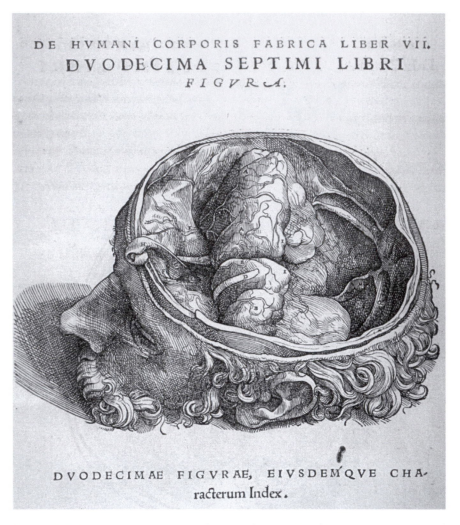

DE HVMANI CORPORIS FABRICA LIBER VII.
DVODECIMA SEPTIMI LIBRI
FIGVRA.

DVODECIMAE FIGVRAE, EIVSDEMQVE CHA-
racterum Index.

Figure 2.4. Even in presenting a human brain, Vesalius is unconventional and provides a personal encounter to the viewer. (Photo: National Library of Medicine)

and Michael Servetus, a Spanish physician. For Servetus this was a costly scientific endeavor and he was burned at the stake for heresy due to his violation of religious actions. Voltaire later defended Servetus as a victim of ecclesiastical totalitarianism (Hillar, 1997).

Girolamo Fabrizzi, known by his Latin name of Hieronymus Fabricius, came to occupy the chair of anatomy at Padua once occupied by Vesalius. He continued the anatomical performances, drawing spectators from outside the academy to his demonstrations held in the University of Padua's second permanent anatomy theater (Klestinec, 2004). The actual dissection of the human corpse typically was informally performed in a more private chamber and completed before the demonstration began in the theater. The formal demonstration was preceded by a formal academic processional, accompanied by musicians, and then interspersed with philosophical instruction.

William Harvey, who had traveled to Padua to learn anatomy in the seventeenth century, practiced dissections using a variety of animals and further developed our understanding of cardiovascular anatomy and physiology. Much of Harvey's work still stands today as remarkably accurate. Later, Harvey also dissected his own father and sister following their natural deaths (Richardson, 2000). Aside from these controversial acts, Harvey's courageous approach set the stage for modern medical research.

Part of the suspense and excitement of early dissections stemmed from their being special occasions held during the carnival season, because this was when cadavers were more available. Apparently, the raucous and sometimes dangerous celebrations allowed thieves and murderers to run wild, resulting in an enhanced supply of corpses. The dissection theater often was a temporary structure built for the purpose. Although observers probably knew where the cadavers came from, they may not have known that they were witnessing new discoveries about the human body.

Dissection became an informative tool to systematically investigate human ailments in France in the eighteenth century as the corpses of people dying in public hospitals were subjected to postmortem dissections. The observations could be related to the signs and symptoms the people were experiencing while alive. Scientists paid particular attention to birth deformities as they tried to understand aspects of circulation and respiration. Looking at the birth defects (teratology) sometimes proved informative, as with a case of Jean Mery examining Siamese twins whose deaths were caused by entangling of their umbilical cords. He demonstrated that oxygen was supplied to the unborn through the umbilical cord (cited in Moscoso, 1998).

DISSECTION AND ANATOMY IN MEDICAL EDUCATION

The dissection theater and the anatomical museum are two essential architectural structures that remain today although in an updated style. The earliest mention of the construction of an anatomical theatre was one made of wood, described by the sixteenth century Spanish monk, Benedictus (Cunningham, 1997).

Dissection Theaters

Conducting a dissection in such a way that a number of students crowded around could observe and learn required a suitable setting that provided adequate viewing. The solution was a theater modeled on a theater for plays, with elevated rows for onlookers. In the demonstrations by Vesalius and others, the lecture and accompanying demonstration-dissection assumed a ceremonial aspect, likened to a blend of a judicial court, a stage drama, and a religious ceremony (Sawday, 1995).

Because preservatives had not yet been discovered, dissections were conducted during the cold winter to avoid putrefaction of the cadaver (Cunningham, 1997). Such theatres of dissection were constructed in metropolitan centers such as Paris, London, Amsterdam, Rome, as well as at medical schools such as Oxford, Leiden, Montpelier, Padua, and Bologna in the sixteenth and seventeenth centuries, and could be fashionable places in which to see and be seen (Sawday, 1995; French, 1999). Public dissections by Vesalius took place inside a temporary wooden theater designed by Serlio. A new structure was erected every time an anatomy lesson was held, whether in Bologna or Padua, sufficient in size that seventy or eighty people could be crowded into the space of the theater. For the famous demonstration by Vesalius of 1540 in Bologna, in his collaboration with Matteo Corti, at least 500 people were present. Particular roles were assigned to the people involved, rather like a theatrical production. The first permanent anatomical theater was built at the University of Padua in 1594 (Carlino, 1999).

Indicating the popularity and importance of the occasion of dissection, once it became established as a teaching method, artists painted the instructional scene of dissecting the cadaver, as with two paintings by Rembrandt (Figures 2.5a–b) and subsequently portrayed by Hogarth (Figure 2.6).

Museums and Preserved Specimens

The anatomical museum typically included a collection of specimens acquired over time, plus models, molds, and other teaching resources that could assist students in learning anatomy. Extensive anatomical museums attached to a lecture theater are still important requirements for a strong instructional program in anatomy (Spence and Zuckerman, 1967). One of the original museums remains today at the Royal College of Surgeons in London and it still is one of the greatest collections of embalmed preparations in all of Europe. The technique for preserving specimens was discovered in the 1660s by Robert Boyle (Asma, 2001). Anatomical museums that included bizarre congenital birth defects and other diseased specimens were popular in Europe, and then New York in the 1840s, and later in other major U.S. cities (Sappol, 2002). The introduction of wax figures allowed bodies to appear extremely realistic. They were featured in some anatomical museums, much as they are today in popular commercial wax museums. Plastination today is starting to replace the less permanent wax figures. Plastination of human figures differs from wax figures in that the process uses actual human bodies that are legally willed to the collection.

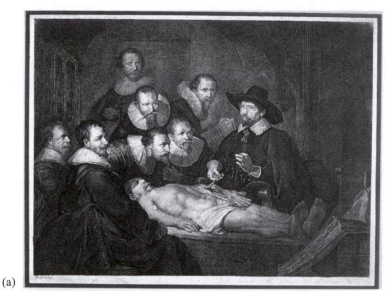

(a)

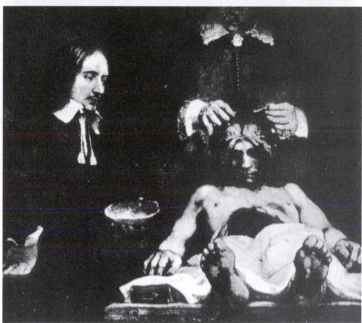

(b)

Figure 2.5a–b. For Rembrandt, the oil painting of the Anatomy Lesson of Dr. Nicolaes Tulp (1632) launched his career and fame. An engraved version of the image is shown here. The oil painting shows Dr. Tulp dissecting the arm of the corpse (*top*). The public dissection was held in 1632 following a hanging (Ricketts, 2006). In reality, however, it would have been unlikely to dissect the arm prior to opening the abdomen. Following another hanging, The Anatomy Lesson of Dr. Jan Deyman (1656) (*bottom*) depicts a dissection of the brain conducted following the emptying of the abdomen. An assistant holds the frontal lobes of the brain in his hand. Both paintings convey a message on the seriousness of death (Ricketts, 2006). (Photos: National Library of Medicine)

Figure 2.6. In 1751, William Hogarth published a series of four engravings on The Four Stages of Cruelty (Jaffe, 2003). In the fourth
engraving, a fictional character named Tom Nero has been cut down
from the gallows and is being dissected in a gruesome portrayal. In the
first print, Tom had been shown torturing small animals, and in the
second, he was beating his horse. The caption on print four refers to
the unthinkable disgrace of Tom's plight in being deprived of a conventional burial, "Behold the villain's dire disgrace Not death itself can
end He finds no peaceful burial place . . . " (Photo: National Library of
Medicine)

Where Cadavers Come From

Acquiring a supply of fresh material for dissection was an ongoing problem, especially prior to the advent of preservation techniques. A profound
abhorrence of mutilating bodies stretches across cultures and centuries. In

the *Iliad*, the earliest literary work of the western world, Achilles drags the body of his enemy Hector around Troy in order to mutilate it and thus dishonor and torment the Trojans. Christians faced an additional issue of requiring the body to be whole during the Resurrection. During the seventeenth and eighteenth centuries, murderers were given an abhorrent additional, postmortem punishment of dissection, promising naked dismemberment, a fate worse than death (Richardson, 2006). Even today, many people are resistant to autopsy, and mutilation of a body is a deep political and personal insult in warfare or terrorism. These deep feelings have always been interwoven with the controversy of performing dissection and increased the challenge of obtaining cadavers.

Once obtained, cadavers would quickly decompose, so the cooler winter became the anatomy season for demonstrations to medical students, when there was a type of natural refrigeration. The selected season also corresponded to Carnival, when a more frolicking atmosphere prevailed, contributing to the unruliness that, while increasing the supply of cadavers, sometimes disrupted dissections.

Because a constant supply of cadavers was required each year, a side issue was that the practice of dissection was riddled with questionable deals among the clergy and the physicians associated with the university, for which cadavers were supplied. Which bodies were appropriate to dissect, which people would participate in performing the dissection, and who would be appropriate observers—all had to be specified. Criminals who had been sentenced to be hanged could be given an additional punishment of being dissected, adding humiliation and depriving the criminal of a proper burial in sanctified ground. Legislation was passed in 1752 in England requiring that all murderers be dissected, as an additional punishment for their crimes (Marshall, 1995), thus providing a primary source of cadavers. Following the execution, the body was conveyed by the sheriff to the Surgeons Company, to be dissected, as illustrated by this passage from the civil code, "In no case whatsoever the body of any murderer shall be suffered to be buried; unless after such body shall have been dissected and anatomized as aforesaid" (Sawday, 1995, p. 55). Clearly, the denial of burial was a posthumous punishment of the criminal's soul.

By 1800, the increasing practice of dissection and the subsequent shortage of cadavers drove up the cost of cadavers. The demand for corpses then led to a brisk business of grave robbing and sale of corpses to dissectors with a thousand corpses a year disappearing from burial grounds in England and Scotland. Grave robbers and their friends were sometimes referred to as "resurrectionists". Grave robbing was focused mostly on fresh graves where the body was less decomposed and the loose soil made easy digging, and thus recently buried bodies were the most at risk.

Perhaps the most notorious case of criminal acquisition of cadavers was seen in 1828, in the Burke and Hare scandal in Edinburgh. Burke and Hare killed destitute people to obtain a source of bodies, smothering them to leave no sign on the body of violence. Over time they committed sixteen murders and sold the bodies to a surgeon, Robert Knox, known as a brilliant teacher of anatomy. The Edinburgh street rhyme of the day was, "Burke's the butcher, Hare's the thief, Knox the boy who buys the beef" (Marshall, 1995, p. 1). Burke was hanged and dissected, and a new verb was coined for murder, to "burke". Knox remained unpunished for his role in the notorious Burke and Hare affair, but numerous surgeons continued relying upon this type of subterfuge to obtain cadavers, especially bodies with deformities (Richardson, 2000). It was even rumored that students could provide cadavers to cover their tuition fees for the dissecting course.

In the midst of all this grave robbing turmoil, the Royal College of Surgeons continued lobbying for a steady supply of bodies to be supplied to surgeons. The surgeon John Abernethy proposed that bodies of persons dying in hospitals, poor-houses, and prisons, and unclaimed by relatives, be given to the surgeons for dissection (Marshall, 1995, p. 9). This proposal was adopted in the Anatomy Act in 1832. The law provided that the government should confiscate the bodies of paupers who died in workhouses and hospitals and could not afford funerals. Yet, even as late as 1880 to 1901, false funerals were held when paupers died, and their bodies sold for dissection and dismemberment at the Cambridge Anatomical Teaching School (Hurren, 2004). Developments such as this policy eventually brought an end to the problem of grave robbing in England (Richardson, 2000), but created enormous distress among the dying destitute.

In America, anatomy became the essential foundation of medical school instruction. Here, as in England, shortage of cadavers led to dissection of executed people's bodies and plundering of graves (Figure 2.7). Body snatching also arose in the United States; a list of grave robbing examples during 1878 includes mention even of many midwestern cities where such incidents occurred (Sappol, 2002, p. 317). This in turn provoked riots showing public indignation at this practice. Legislation was passed in many states allowing the bodies of the indigent to be given to medical schools for dissection (Sappol, 2002, pp. 123–124).

Medical schools expanded rapidly in the states as exemplified by the example of growing from 4 schools in 1800 to over 160 in 1900. Concurrently, questions related to the use of bodies came to involve philosophical, psychological, sociological, anthropological, and medical discussions of the self and body (Ratzan, 2004). Considering the complexity of the issues, we should not be surprised to learn that the topic profoundly affected people's emotions and interest, leading occasionally to riots and widespread fear of bodysnatching among the dying indigent. At the same time, however, anatomical spectacles included public lectures on anatomy that attracted huge audiences, popular anatomical books with large readerships, and popular anatomical museums. Of the popular picture books and textbooks on anatomy, the 1837 book by William Andrus Alcott, *The House I Live in*,

Figure 2.7. The shortage of human bodies for dissection led to extreme measures, such as digging up bodies from fresh graves, and led to unsavory financial arrangements to obtain bodies. (Photo: National Library of Medicine)

was perhaps the most favored. This book was made popular by missionaries who translated it into other languages (Sappol, 2002, 2003).

By the eighteenth century, anatomy had become a topic of study at major universities in Europe. Edinburgh and London stand out in this regard. Two brothers in London, surgeon John Hunter and physician William Hunter, provided noted leadership in their "anatomy schools" in London, offering skillful technique and the resources of John Hunter's anatomical collection (Sappol, 2002). Private medical and anatomy schools proliferated, offering outstanding instruction with anatomy as the centerpiece, including personal opportunities for dissection and exposure to detailed gross and comparative anatomy, as well as unusual specimens. Although eventually the nefarious dealings with grave robbing and body snatching came to an end, the investigations of human and animal anatomy continued to arouse antipathy toward some famous figures who today are honored for their wisdom and contributions.

Religious prohibitions against defiling bodies helped fuel controversies in the use of animals for dissection. Restrictions by the Jewish and Muslim religions, and Christian expectations for resurrection of the whole body continued to fuel the controversy. As noted, Vesalius is honored by scientists for his pioneering work on the circulatory system and his seven-volume contribution certainly is the most influential medical work ever (accessible on the Web: vesalius.northwestern.edu). However, he was charged with a crime at the time for this contribution and sentenced to death by the Inquisition. Eventually, his sentence was commuted with a requirement that he undertake a pilgrimage to Jerusalem to expiate his guilt.

This brief, if sometimes macabre and complex history of dissection and anatomy, comprises the development of medicine as we know it. Anatomy, both human and animal, is the mainstay of biology. For evidence of this one needs only consider that *Gray's Anatomy* is undoubtedly the most well-known medical textbook in the United States. The name is such a household word that one could argue the term "Gray's Anatomy" opened the door for a very successful television series of medical dramas.

History illustrates that the philosophical and ethical issues surrounding dissection of human and animal bodies were a long-standing concern, and portrays the advancement of knowledge from a naivete about even the most basic knowledge about human bodies, through the impacts and gains that we can attribute to dissection. As we move toward the realization that knowledge gained by dissection is vital to advances in human and animal health, we are beginning to realize that perhaps there should be a line drawn between the appropriate educational and research interest, and the use of dissection simply because this has been done as the standard practice in the past.

CHAPTER 3

Controversy in the Development of Science Education and Dissection

From looking at the early history of dissection in Chapter 2, we have seen that the presentation of dissection as a tool for learning biology has been modeled on the ancient practice for acquiring new scientific knowledge and for teaching medical knowledge and surgical skills, especially to medical students. In this tradition, both human and animal cadavers were used to teach students about human biology. In this chapter we will consider, how that practice of dissection made its way into a convention which is used not only in medical professional education, but has become commonplace for precollege children. In colonial America and into the nineteenth century, reading, writing, and arithmetic were the basics of childhood education. Being literate was a mark of being well-educated. Many children had to leave school around eighth grade, or even before gaining competent reading and writing skills, in order to work on their family farms or otherwise help their families. Instruction in the early grades focused on the basic skills. Classical topics such as languages and literature were emphasized in the secondary levels for those children who remained in school long enough for more extensive education.

In the nineteenth century, biology as an educational subject was introduced at the precollege level. Once these curricula became more widespread, philosophers and psychologists subsequently engaged in an ongoing dialog concerning how much biology should be taught and from what perspective. Social movements arose—behaviorism, pragmatism, and constructivism—and each had a direct impact on science education, as described in this chapter. Political concerns, such as that triggered by the launch of Sputnik in 1957, at times have prevailed in creating awareness of major educational deficiencies. This led to rapid shifts in the direction of science education.

Biology education, which became pretty much widespread around the time of Darwin, has been buffeted by many forces outside the classroom. DeBoer (1991) wrote a central resource in examining this period of the history of science education.

BIOLOGY INSTRUCTION IN THE NINETEENTH CENTURY

During the late eighteenth and early nineteenth century, many philosophers and educational theorists advocated that children should be able to make discoveries for themselves, rather than being bored with rote learning and recitation. Conceptual contributions were built on the work in Europe of Jean Jacques Rousseau in the eighteenth century, as well as Johann Heinrich Pestalozzi and Johann Friedrich Herbart in the nineteenth century. While many of these pioneering thinkers reacted against telling students what they needed to learn and wanted students to make their own discoveries, Herbart (1835/1901) advanced an educational theory of four steps in the process by which children form new concepts. In brief, these are: (1) present ideas related to the experiences and knowledge the child has acquired already; (2) use conversation to allow the student to develop the new understanding in the most convenient way; (3) then use directed instruction to teach something the student cannot learn alone; and (4) finally, ask the student to apply the knowledge in a new problem-solving manner. With these forward thinking concepts one can see why Herbart is also credited with emphasizing being practical in helping children become fit for life.

While conventional grammar schools continued to emphasize classics through the nineteenth century, a broader curriculum evolved in the United States within the new academies from about 1750 to 1850 (DeBoer, 1991). It offered utilitarian coursework geared toward a practical education that included a wide range of subjects such as botany, chemistry, and astronomy, though without offering any practical or laboratory experience (Sizer, 1964). The academies lacked expertise among their faculty, and had poor teaching materials. Often the courses were short and poorly organized. Nonetheless, the academies offered science courses that were not overshadowed by the classics. For the first time, anatomy was popularized for the reading public, especially the 1837 book by William Andrus Alcott, *The House I Live in* (Sappol, 2003). Missionaries published not only the *Bible* and other religious tracts, but also favored this book. Working with the Karen people of Burma, the missionaries hoped the study of anatomy would get them to give up their native ideas about the body. They translated the book into other languages, with anatomical illustrations that supported a religious emphasis, as illustrated in Figure 3.1 (Sappol, 2002, 2003).

What steps led toward adding science into the classroom? How were science laboratories introduced into precollege education? When we think

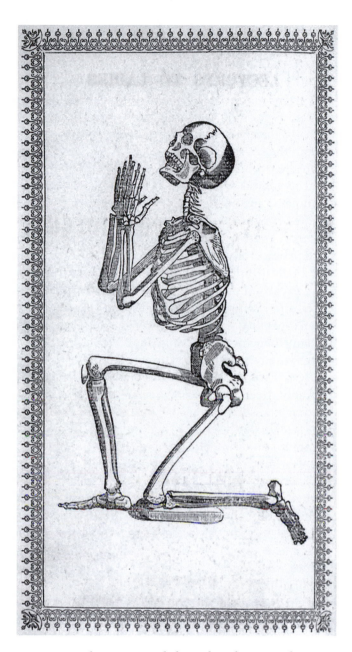

Figure 3.1. This praying skeleton based on one drawn in 1713, became a feature of William A. Alcott's popular anatomy book, *The House I Live In* (1832) that was popular among missionaries who carried it to foreign countries. It was also used in Mary Gove's *Lectures to Ladies on Anatomy and Physiology* (1842) to assure their moral improvement. (Photo: National Library of Medicine)

of the nineteenth century, the explosion of scientific discoveries immediately comes to mind. Charles Darwin (1859/2003) changed the paradigm of how animals (and, presumably, humans) were viewed with his book, *The Origin of Species by Means of Natural Selection*. His work, and that of others, brought science into the realm of everyone, raising practical questions that impacted views of everyday living. Philosophical discussions began to emphasize topics of science and culture, reviving topics that had been prominent in the days of Hippocrates and Aristotle in ancient Greece. Following up on the evolutionary ideas of Darwin, Thomas Huxley promoted the anthropological perspective. He considered the implications of evolution for humans, and addressed its relationship to education and culture. Some themes of Huxley's work were echoed in the work of his three grandsons, Aldous, Andrew, and Julian, who helped retain Huxley as a household word in academic communities, as, of course, was Darwin.

Advocates of modernization of education argued for a shift in perspective to incorporate science to address practical social problems such as sanitation, poverty, and education, whereas classicists argued that the development of one's intellectual capability was best achieved by studying ancient works of Greek and Roman civilizations and languages (DeBoer, 1991). Among the host of advocates who campaigned in the nineteenth century for teaching science to children, Thomas Huxley's writings remain available and well known. Remarkably, traces of the controversies concerning how to regard Darwin's work remain as lively topics of discussion today.

Some of Thomas Huxley's writings address topics that still seem fresh. A strong advocate for studying biology and learning about one's body, he argued that people's knowledge of germ theory and of practical measures to avoid disease was important as such knowledge would help people maintain good health (Huxley, 1854/1902). Since science could help develop the mind and stimulate one's interactions with the natural world, Huxley (1876/1902) favored introducing science, especially biology, to children at as early an age as possible. He strongly believed in the importance of interactive science, such as handling the plant or organic material being studied, and feeling the physical forces in a class on gravity or centripetal forces. He set the stage for establishing classroom laboratories, and argued that reading alone, without the direct experience to provide images of nature, would not provide the important understandings.

In a series of essays and addresses, Huxley became a tireless advocate for the value of childhood science education. In his 1854 paper, "On the educational value of the natural history sciences," he argued that the methods of biological science are identical to those of the physical sciences, including observations of experiments, comparison and classification of similar facts, deduction, and verification (in Huxley, 1854/1902, p. 51). He believed that the common facts of biology may be taught with advantage to the youngest child (Ibid., pp. 61–62). His 1869 paper, "Scientific education," suggested that no boy or girl should leave school without knowing the general character of science, and without being disciplined in its methods (in Huxley, 1869/1902, p. 110). He pointed out that the teacher should really know his subject, and speak of it in everyday easy language.

The term, *biology*, with its concept of uniting the sciences that deal with living matter into one discipline, only arose around 1800, replacing the term *natural history*. As described by Huxley in his 1876 paper, "On the study of biology" (Huxley, 1876/1902, p. 228), Bichat, Lamarck, and Treviranus almost simultaneously put forward the concept of this new discipline.

Huxley (1877/1902, pp. 247–248) described the biological laboratory where he lectured, and highlighted dissection, mentioning a room for practical work, with tables, microscopes, and dissecting instruments for dissecting the structure of a certain number of animals and plants (p. 248). He advocated learning from your own body.

While arguing for instruction in physiology based upon observation, he also noted the contradiction of the then-common practice of boys fishing with stabbed live frog bait, yet not being allowed to study a frog in the classroom ("On elementary instruction in physiology," Huxley, 1877/1902, pp. 255–256).

Perhaps less well-known, in the late nineteenth century Herbert Spencer became another strong advocate of providing children with opportunities to make their own investigations and draw their own inferences, allowing them to discover as much as possible (Spencer, 1864). He noticed that children constantly and naturally go through processes of observation, inquiry, and inference, and argued for their self-development. He emphasized that telling children what they should learn, rather than allowing them to direct their own learning, could make learning aversive to them.

Science education continued to expand in classrooms through the United States, as illustrated in Figures 3.2a–c. By the late nineteenth century, a report on secondary schools prepared by the National Education Association's Committee of Ten (1893) led to the standardization of public school curricula, with strong leadership provided by academic leaders from several institutions. This group advocated a heavy emphasis on natural science and scientific inquiry, with double laboratory periods and extensive field trips (Mintzes and Wandersee, 1997a).

SCIENCE EDUCATION IN THE TWENTIETH CENTURY

Since the launch of science education on a widespread basis, its objectives have mirrored the issues prevalent in broader society. During the twentieth century, the professional fields of psychology and education each developed their philosophical and research frameworks. Over the same period, the major objective of education, and especially science education, continually evolved, with the primary objectives changing in ways that affected science classrooms. Although the knowledge base concerning the learning process continued to develop, the emphases for science and health education can

(a)

(b)

Figure 3.2a–c. During the 19th century, the study of anatomy and health became widespread for students in precollege through medical school classrooms. These images show (*top to bottom*): students in Washington, DC; a dissection conducted in a women's medical college in New York City; and a similar lesson around the same time in Sudan, Africa. (Photos: Library of Congress)

(c)

Figure 3.2. (*continued*)

be characterized as a see-saw pattern of shifting back and forth from an emphasis on students learning an elitist body of established knowledge and fulfilling a set of corollary testing requirements, to a view of all learners creating their own learning structure and a body of knowledge that is important to them for everyday living. Changes in the educational goals affected the presentation of science and the expectations for preparing students in scientific knowledge. Throughout the century, the available amount of knowledge of biology and the methods and technology for its study have exploded exponentially, creating further quandaries for science educators. What is science, and what should children be taught about it? This topic is continually revisited and reassessed by society at large, impacting the expectations and objectives set for teachers in their science classrooms. Specific requirements associated with these expectations can include mandatory testing of students, with budgetary consequences.

Pragmatic Progressivism

At the beginning of the twentieth century, John Dewey (1916, 1938) and other educators emphasized the practical learning that students would take into their everyday lives. Termed the progressive movement, this period emphasized usefulness of scientific knowledge and vocational training. These forces converged in a report of the Commission on the Reorganization of Secondary Education (National Education Association, 1918) that recommended a curriculum that would prepare students for the activities of life, including health, knowledge of fundamental subject matter, family life, vocation, citizenship, worthwhile leisure, and ethical character (Mintzes and Wandersee, 1997a). During the same years, new methods of educational testing were developed by Edward Thorndike and Louis Terman, focused on assessing the efficiency of the learning process. Montessori's (1912) methods of student-initiated learning acknowledged the students' participation in the learning process, choosing what to learn and the timing of the learning.

Also contributing to this progressive period, an educational approach termed behaviorism evolved from research in operant conditioning, an addition to educational learning theory. The behavior of rats and mice could be manipulated and shaped through so-called stimulus-response, rewarding them for accurate learning. Movements such as behaviorism within the developing field of psychology inevitably impacted education. The introduction of behaviorism led to the development of techniques for behavior modification, initially associated with experimental work on operant conditioning, and came to be applied to the educational setting. This movement in psychology had wide-ranging influences on child rearing practices and educational approaches. It drew from research on systems of learning, spearheaded by B. F. Skinner's (1938) study of rats and pigeons, incorporating the concepts of operant conditioning and classical conditioning, and reinforcing the perspective that the behavior of students could be shaped and molded. The concept of conditioning was employed in developing techniques for teaching children as it was for training pets.

Sputniked Back to Subject Matter

A major wake up call resounded through the United States during the Cold War when Russia launched the first artificial satellite, Sputnik, on October 4, 1957, preceding America in this accomplishment. Suddenly, the state of science education was viewed as a national crisis, and political attention was mobilized for a shift in direction. Science education "was itself co-opted to serve the dominant social and political interests of the time" (Rudolph, 2002, p. 116). The changing state of science education continued to be a focus of attention from that time (Library of Congress, 1990). Major initiatives for science curricular reform funded by the National Science Foundation arose from inside academic circles, including the NSF Biological Sciences Curriculum Study (BSCS) project. The reforms following Sputnik stressed the nature of scientific inquiry, while at the same time, the educational community became more sophisticated in conducting and publishing research (Mintzes and Wandersee, 1997a, b).

By the early 1960s, scientists had generated a public policy consensus and initiated curriculum reform, spearheading several projects, including

BSCS that was headed by Bentley Glass and Arnold Grobman at the University of Colorado in Boulder (Rudolph, 2002). BSCS, the first of the federally-funded curriculum projects, sought to convey both the richness and explanatory power of Darwin's ideas, presented the theory of evolution by natural selection, and became the model for other projects. The country's top research scientists were mobilized into producing educational materials. Along with solid science, they included applied information that would be practical for the K-12 students, whereas more theoretical and abstract concepts were considered as appropriate for the college level. The textbooks included analyses of pressing social problems from a biological perspective, examining problems of modern man such as the dangers of radiation, pollution, overpopulation, and the deterioration of natural resources. These National Science Foundation-funded curriculum projects were discipline-centered and paid explicit attention to the nature of science. Such curricula downplayed the curriculum emphasizing life adjustment and preparation stemming from the progressive education movement. They valued *science inquiry*, and encouraged students to ask questions, make observations, record and process data, and develop conclusions (Biological Sciences Curriculum Study: BSCS, 1963). While they created great excitement at that time, these curricula have fallen into disuse, and are viewed as too difficult, elitist, or outdated (Rudolph, 2002).

Constructivism

The next period in the development of science education was less dramatic than the Sputnik occasion, occurring as a broad movement toward experimentation, collaboration, inquiry-based learning focusing on the process as well as content, and the understanding that humans are, in fact, animals. The pendulum inevitably swung back to acknowledge a less concrete view of science and reality, and a more active view of the learner. The high impact of *The Structure of Scientific Revolutions* (Kuhn, 1962) rippled through education, conveying the postmodernist notion that science resides within the culture, and leading to some challenging questions concerning the objectivity of science (Matthews, 2000).

This movement built on, yet differed from, theories of child development by Jean Piaget and cognitive psychology by David Ausubel. Ausubel (1968) had made a distinction between rote learning and meaningful learning that is incorporated into a learners' framework of knowledge, or cognitive structure. In a major paradigm change termed constructivism, the process by which students learn science came to be identified with "constructing personal meaning." This movement in 1980, and continuing into the 1990s, was linked with postmodernist and antirealist views about the nature of science (Matthews, 2000). An example is the claim that "the authority for truth lies with each of us" (Driver and Bell, 1986, p. 452).

Constructivism became a major influence in education in the late twentieth century, despite there being some perils of too closely following its tenets (Matthews, 1998). While there is essential value in understanding the effects of culture in shaping perspectives and knowledge, a key question concerns how constructivist principles come to bear in teaching abstract scientific principles (Matthews, 1998, p. 8). How does one lead students to construct for themselves large bodies of knowledge that already are organized as major systems?

The loss of interest in science during the 1990s has by now become a major concern (Matthews, 2000). The flight from science, especially at the college level, seems widespread, not only in the United States, but in Europe and Australia as well. According to one authority, "If children have no patience for learning, if the bulk of their time is spent watching brain-dead television, then the idea of 'science for all' is fanciful. The problems of science education cannot be divorced from the problem of education in general" (Matthews, 2000, p. 8). The goal of offering science to all, and achieving widespread science literacy is hampered by the widespread inadequacy of adults' skills in mathematics and reading (*Adult Literacy in America*, U.S. Department of Education: Kirsch, Jungeblut, Jenkins, and Kolstad, 1993).

Social Issues as Priorities

Philosophical considerations based on the ideas of Conant (1947) and Kuhn (1962) criticized the positivists' position that science is a neutral idea, that scientific observation is based on theory, and subsequently explored basic epistemological questions of how we come to know the external world (Mintzes and Wandersee, 1997a). Following on the constructivist emphasis on the learner, a movement was underway emphasizing the importance of science literacy for all. Multiculturalism and gender equity came sharply to the center of attention. The need to integrate practical applications, and to incorporate new technologies and mathematics led to the acronym on science, technology, and mathematics, STEM education, in the 1980s.

Influences of educational research on how children learn are only one element of the initiatives for science educational reform. Political forces give prominence to current emphases in education on diversity and equal access. Particularly in certain states, conflicts center on presentation of material on evolution versus intelligent design.

Human Constructivism

Sometimes it is easiest to understand an innovation by seeing it work in one person's life. For example, the pioneering educator, Joseph Novak (1997), has described his childhood experience of struggling to find books that explained how things worked and prodded readers to reason things out. His hunger for a different type of learning philosophy was satisfied by the work of Conant (1947) and Kuhn (1962), which helped him formulate advancing constructivism. His work focused on gains in problem-solving ability related to acquiring better organized knowledge structures. Through struggling with various paradigms, Novak created a theory of education specifying:

1. *Meaningful* learning employs well-organized knowledge structures.
2. Along with hands-on experience, integration and careful clarification of words are important.
3. Approaches to learning vary, and are somewhat related to learners' epistemological ideas.

Of particular interest here, Novak conducted a longitudinal study of science learning, creating audiotutorials to investigate the effects of self-paced study by students, and assessing the individual strategies used by students.

As an indication of the emphasis on the human approach, the National Science Teachers Association has assessed how recent science curricula incorporate the key findings about how people learn into the science materials development. Key findings include, that students arrive in class with preconceptions; that students need to develop factual understanding based in a conceptual framework; and they need to set goals and analyze their progress toward them (Bybee, 2002). One example used as an illustration is the BSCS 5E Instructional Model: engage, explore, explain, elaborate, and evaluate.

Expanding Learning Opportunities

Even as science in classrooms has been buffeted back and forth by international and national politics, the presentation of science has arisen in new venues. Some of the most effective opportunities for learning biology today arise outside the classroom, as presented in a variety of informal science education settings, or in home schools.

Informal Science Education

Almost everyone is well aware of the popular science educational programs that attract wide audiences and yet manage to present scientifically accurate and often cutting edge material and do it in an appealing way. Nature programs on the Disney Channel, Animal Planet, and PBS's Nova are examples.

Another expanding area of the democratization of learning is represented in the informal science education movement. Everyone now has the opportunity for lifelong learning involving exploration of the world, not only at zoos, but by using the array of technical electronic methods, or visiting engaging science museums and educational displays. The Exploratorium in San Francisco is both a museum and display of cutting-edge science and art that attracts hundreds of thousands of visitors each year (Delacote, 1998). In what is termed a "learner-centric" approach, high school and college students are employed to become explainers at the museum, and help visitors enjoy the exhibits in the museum. This approach has proven valuable to both the visitors and the explainers, who learn by teaching. Museums like the Exploratorium dramatize a culture of learning, bringing multiplex, interdisciplinary, interactive experiences to visitors (Delacote, 1998). Even many smaller communities feature science museums, such as Explorit, in Davis, California, that serve thousands of children who visit each year (Figure 3.3).

The Advent of Home Schooling

A growing number of parents are choosing to home-school their children, furthering a growing industry of science materials designed for this use. They make extensive use of local resources such as science museums. One study of over 20,000 home-schooled students reported that their achievement test scores were typically in the seventieth to eightieth percentile (Rudner, 1999). When enrolling in public school, 25 percent were placed one or more grades above their age-level.

CONTENTION OVER DISSECTION IN EDUCATION

For biology classes in intermediate and high schools, dissection has been almost an iconic symbol of the course. The practice has become so traditional that only with some difficulty or major initiative will it be disrupted. Although the topic of dissection rarely appears in current precollege educational literature, it continues to be a focus of attention among animal protection organizations. At one period in the early 1990s, the topic of dissection erupted as a storm center in which leading science educators and the National Association of Biology Teachers engaged. A concise summary of some of the key moves in this drama appeared in the paper by Joseph McInerney (1993) in *The American Biology Teacher*, "Animals in education: Are we prisoners of false sentiment?" Reflecting his perspective, the paper asks if our use of animals is less than what it could be, because of our sentimentality towards animals. The author was then director of the Biological Sciences Curriculum Study, which created outstanding curricular materials. During this series of altercations, the National Association of Biology Teachers issued a policy statement supporting scaling down dissection, which aroused an uproar among some educators and scientists, and the policy then was rewritten and clarified in a series of skirmishes and published critiques. For all the contention during this period, nothing productive seemed to come out of it except some unpleasant history.

Figure 3.3. Science is a hands-on endeavor at science centers for children, such as the Explorit Science Center, where this young visitor faces up to a replica of a dinosaur skull. Local science centers create revolving exhibits, providing hands-on access to science for children who visit. Explorit is just one of a network of 250 science centers in the Association of Science and Technology Centers network. Explorit engages over 70,000 people from 13 California counties per year, and serves a diverse community of families and school groups. [Courtesy of Neil Michel/AxiomPhoto.com]

Conflict and Assessment Regarding Dissection

A combination of criticism, cost, and inconvenience, paired with the emergence of interests in cellular, genetic, and molecular emphases, often leads toward reduced efforts to offer meaningful laboratories on whole animals to students. This trend can be seen in both K-12 and college education, leading to concerns that the study of anatomy, and knowledgeable scholars who could teach it, are disappearing.

Laboratory Experiments

Laboratory classes in precollege biology offer students the advantage of getting hands-on experience in an interactive manner. In this process, students acquire laboratory skills and experience in handling data. In sometimes partnering with the teacher in exploring the laboratory materials, they become more acquainted, or even a little bonded, with the laboratory teachers. Despite the tradition of laboratory-based biology courses across the United States, such courses are being phased out in favor of more molecular-level studies. Fewer students graduate with expertise in dealing with whole animals, to the point that alarm is sometimes expressed (e.g., Janda, 2002). A further problem has arisen at medical schools, where qualified instructors for teaching gross anatomy are in short supply, another consequence of anatomy becoming increasingly molecular and focusing on cells rather than organs (Holden, 2003a).

Over the years, dissection has been practiced, beginning in intermediate grades and continuing in undergraduate and professional education. Its use in precollege education has been discussed from the standpoints of studying attitudes of students to the dissection and criticizing teachers for not shifting over to alternatives that supplant uses of animals.

A leader in the arguments against the practice of dissection, Jonathan Balcombe (2000, 2001, 2002), has provided a comprehensive, ongoing, and reasoned overview of the evolving situation. The Humane Society of the United States (2007b), initially assisted by Balcombe, has focused attention on methods of encouraging and supporting students who prefer not to dissect, and provided them with tools for refusing to participate in the practice. Some major points critics of dissection have made on this topic are that:

1. many alternatives to animals are now available for teaching
2. learning with alternatives is as effective as with animals
3. some alternatives can be borrowed at no cost from lending libraries for alternatives
4. students' attitudes to dissection vary, including both extremes
5. legislation gives a growing number of students the option of refusing to dissect and being provided an alternative method of learning

It may also be instructive to note problems related to the introduction of alternatives that were highlighted in a report on the use of animals in higher education. The workshop panel analyzed the reasons for the small decline in use of animals in education as follows (Van der Valk et al., 1999):

1. resistance of teachers to change
2. initial investment of time and money required
3. poor dissemination of information about potential alternatives
4. varying quality of material available
5. financial, technical, and other factors restricting use of alternatives

Many individuals writing about dissection represent a philosophical perspective. Although the subject has been a lively focus of articles among animal welfare organizations and philosophers, educators have had much less involvement in addressing this question than one might expect. One university professor, Charles Ralph (1996), sharply criticized the traditional, ingrained belief that all students in biology should have a traditional hands-on laboratory experience. He acknowledged some appealing features of a traditional laboratory, including the aromas of specimens and ambience of instruments and apparatus, the socializing opportunities, the bonding with a good instructor, and the practice of manipulative skills. On the other hand, the electronic alternatives offered learning in a quiet, informationally rich environment, a flexible pace of individualized learning, excellent pictorial or graphical representations, and opportunities of data generation with experimental and procedural options. For the most part, however, writers have not explored the perspectives of teachers who are leading the learning of their students, a topic we introduce in Chapter 7.

Learning Effectiveness with Alternatives: The Computer Age

Numerous studies have assessed the effects of students learning in laboratories using animals versus those using alternative methods, as described more fully in Chapter 8. A recent review examined seventeen controlled, comparative studies assessing learning with alternatives in education (Patronek and Rauch, 2007), but only three early studies involved high school biology students, who performed as well with alternatives as with animals (Fowler and Brosius, 1968; Kinzie, Strauss, and Foss, 1993; Strauss and Kinzie, 1994). An annotated Web summary of thirty-five studies dealing with dissection and other animal use as a learning method in comparison with alternatives methods is available (Humane Society of the United States, 2007c). It provides short descriptions of studies assessing various conventional and alternative teaching methods for student performance in veterinary, medical, undergraduate, and precollege levels.

An early careful review of assessments of technological alternatives to actual dissection in anatomy instruction identified some statistical criticisms in the relevant research. Yet, here too, the results were relatively consistent across the various educational levels and contexts, in that "high-tech," computer-driven, simulations were as effective as actual dissection for teaching biological information, and even the noncomputer, alternative techniques such as video did not yield different achievement outcomes from either high-tech or animal dissection (Zirkel and Zirkel, 1997). One concern expressed was that the high-tech packages had not yet reached their interactive potential for instructional efficacy. Empirical studies reviewed were conducted in medical and veterinary schools, undergraduate education, and high school curricula. This review emphasized that little knowledge was available on attitudes to the alternatives, though the Web site, "Net-Frog," had yielded over 2,000 machine visits per week (Kinzie, Larsen, Burch, and Baker, 1996).

In veterinary education, Tufts University was the first veterinary school to establish an alternative training program allowing students not to be required to dissect animals which had been euthanized for the purpose of dissection (Pavletic, Schwartz, Berg, and Knapp, 1994). Cadavers were obtained from ethical sources, often clients who donated their terminally ill or dead pets that were frozen until used. Students in the class of 1990 selecting the alternative laboratory program were assessed by contacting their employers after they graduated and entered practice, and found to be at par with those completing the conventional program. This pilot effort led to modifying the laboratory program and, eventually, entirely phasing out dissection of purpose-bred animals.

Availability of Alternatives

The number of teaching resources that potentially can supplant animal use continues to grow and access to information about them has become easier. NORINA (2007) catalogs detailed information on each of 3,000+ resources, with purchase information and prices listed, as well as descriptions of the product. Sophisticated point-and-click searching tools are available with a tutorial developed by UC Center for Animal Alternatives (2007), for NORINA, as well as those offered by EURCA (European Resource Centre for Alternatives to Using Animals: 2007), AVAR (the Association of Veterinarians for Animal Rights: 2007), and InterNICHE (International Network for Humane Education: 2007). InterNICHE provides a book detailing 500 products, and backs this up with translated versions and a comprehensive Web site (Jukes and Chiuia, 2003). So it is true that there never have been more resources available. Many of these are expensive, and they require that teachers make a selection among the products, perhaps without trying or even ever seeing the product.

Alternatives can be borrowed. Recognizing the limited budgets and resources of teachers, several organizations now maintain lending libraries of teaching resources, providing a no or low cost source of such materials, a topic we discuss more fully in Chapter 6. Use of these loaned materials, though increasing, is limited, especially considering the huge numbers of teachers within the United States that teach biological science in precollege classrooms. The loans offer an option that perhaps serves only the most organized and highly motivated of teachers. Use of these materials requires ordering the materials well in advance and having the schedule well in hand considerably before the date of the intended laboratories.

Legal Endorsement for Students Electing Alternatives

Much of the effort of the animal protection organizations has gone toward establishing communication and materials for students who may prefer not to participate in dissection, as well as support for legislation guaranteeing these students alternatives to dissection if they so choose. Such legislation is discussed in Chapter 5. A legal case in California led to creating the first students' choice law in the United States, in 1988. Since then, many states have passed similar legislation pertaining to precollege instruction as described in Chapter 5. This legal picture in the United States is the opposite of that in some other countries, where the default policy is that dissection for secondary students is banned, in countries as geographically diverse as Argentina, Norway, and Switzerland.

Contrasting Approaches in Precollege and Higher Education Laboratories

Although the practice of dissection continues in precollege classrooms, uses of animals in college and professional school laboratories have sharply declined across recent decades. Paradoxically perhaps, veterinary schools have led the way in creating instructional software and moving away from the traditional consumable uses of animals in instruction. This dramatic shift in teaching methodology was pushed hard by veterinary students' dissatisfaction with dissection. In response, creative developments were profiled by teaching faculty who were striving for maintaining the quality of instruction while meeting the concerns of the students. A commitment from veterinary school administrators to assign funding for development of the improved alternative resources ushered in a new movement that both met academic requirements and recognized the concerns of students (Hart and Wood, 2004).

CHAPTER 4

The Context for Dissection: Educational Testing

Science education and testing command daily front-page headlines and focus our society's attention on the content being taught in science class-rooms and whether it is being learned. The prescribed content for each grade level, in a program with integrated scope and sequence, has been developed through consensus-building among the wide range of communities involved, and is stated in national and state science education standards and frameworks, as presented in Chapter 5. The magnitude of effort involved in the process of defining standards and frameworks speaks to educators' concerns regarding science education and their responsiveness to the need for improvements. What may be surprising to some readers is that *dissection* does not enter the picture concerning the content to be taught and tested upon; dissection is, in fact, never mentioned in any of the documents pertaining to standards or testing. *Dissection* only becomes relevant to classrooms at the point where the curriculum of daily lesson plans and laboratory methods is being worked out. The testing described in this chapter, and the standards and frameworks presented in Chapter 5, establish the overall structure for the entire science education process prior to college and set the required objectives for science teachers. The testing and standards thus define the foundation of biology and health instruction, and set the context in which *dissection* appears in today's classrooms. Since *dissection* is simply a laboratory method of instruction, and not an educational objective in itself to be examined upon, it will scarcely be mentioned in the remainder of this chapter. Rather, the goal of this chapter is to illustrate the context in which dissection is taught. A variety of resources pertinent to testing and this chapter are presented in Table 4.1.

Table 4.1 Educational Testing

Title and URL	Producer	Content
AAAS Project 2061 www.project2061.org	American Association for the Advancement of Science	Long-term AAAS initiative to advance literacy in Science, Mathematics, and Technology.
ACT http://www.act.org/aap/	ACT, Inc.	The ACT test assesses high school students' general educational development and theoretically their ability to complete college-level work. The multiple-choice tests cover four skill areas: English, mathematics, reading, and science. The writing test, which is optional, measures skill in planning and writing a short essay.
ASCD http://www.ascd.org/	Association for Supervision and Curriculum Development	A community of educators, advocating sound policies and sharing best practices to achieve the success of each learner. Offer broad, multiple perspectives, and address aspects of effective teaching and learning, such as professional development, educational leadership, and capacity building.
ASTC Publications http://www.astc.org/pubs/	Association of Science – Technology Centers	ASTC Publications sell books and periodicals for and about the science center field. Topics range from science center management and exhibit planning to research and evaluation.
CSTA http://www.cascience.org/	California Science Teachers Association	Represents California science educators in every science discipline at every grade level, Kindergarten through University. Organizes and participates in statewide reform initiatives and provides opportunities for members who wish to serve on state advisory committees, including framework, standards, and textbook committees.

(continued)

Table 4.1 (*continued*)

Title and URL	Producer	Content
California DOE http://www.cde.ca. gov/	California Department of Education	Mission is to provide leadership, assistance, oversight, and resources so that every Californian has access to an education that meets world-class standards. Goals and objectives include holding local education agencies accountable for student achievement, building local capacity to enable all students to achieve to state standards, professional development in support of teachers, and the effective use of technology.
California DOE Publications http://www.cde.ca. gov/re/pn/fd/		Free downloads of publications available in PDF, including state standards, frameworks, and other department documents.
California STAR Program http://star.cde.ca.gov/ star2003/index.asp	California Department of Education	The California standards tests in English-language arts, mathematics, science, and history–social science are administered to students in California public schools.
STAR Results http://star.cde.ca.gov/		The 2003 California Standardized Testing and Reporting (STAR) program results for the state, counties, districts, and schools are available at this site.
NABT http://www.nabt.org/	National Association of Biology Teachers	Offers conferences, resources, publications, and workshops.
Position Papers http://www.nabt.org/ sites/S1/index.php?p=26		Position papers include: Administrative Support for Life Science Teachers; Biology Teaching Preparation Standards for Middle & Secondary Teachers; Statement on Teaching Evolution; and Role of Laboratory and Field Instruction in Biology Education.

(*continued*)

Table 4.1 (*continued*)

Title and URL	Producer	Content
NEA on NCLB http://www.nea.org/ esea/index.htm	National Education Association	The NEA offers plans to improve the NCLB Act and discusses ways to influence, fix, and change the law.
NSTA http://www.nsta.org/	National Science Teachers Association	Supports standards-based science education and promotes excellence and innovation in science teaching and learning, curriculum and instruction, and assessment. Promotes interest and support for science education. Supports science teachers, science supervisors, administrators, scientists, and others involved in and committed to science education with journals, publications, and conferences.
A Nation at Risk: the Imperative for Educational Reform http://www.ed.gov/ pubs/NatAtRisk/ index.html	The National Commission on Excellence in Education	Report to the Nation and the Secretary of Education, United States Department of Education by The National Commission on Excellence in Education, 1983, includes findings and recommendations.
National Research Council http://www. nationalacademies.org/ nrc/	National Academies	Part of the National Academies, NRC is a private, nonprofit institution that provides science, technology, and health policy advice under a congressional charter. It was organized to associate the broad community of science and technology with the academy's purposes of further knowledge and advising the federal government, and provides services to the government, the public and the scientific and engineering communities.
National Sciences Education Standards http://www.nap.edu/ readingroom/books/ nses/	National Academies Press	Presents a vision of a scientifically literate populace, outlining what students need to know, understand, and be able to do to be scientifically literate at different grade levels.

(*continued*)

Table 4.1 (*continued*)

Title and URL	Producer	Content
The Nation's Report Card http://nces.ed.gov/ nationsreportcard/	National Assessment of Educational Progress	Offers continuing assessment of what America's students know and can do in various subject areas, including math and science.
No Child Left Behind Public Law http://www.ed.gov/ policy/elsec/leg/esea02/ index.html	US Department of Education	Public Law print of PL 107-110, the No Child Left Behind Act of 2001, Table of Contents, Fulltext pdf and Overview.
NCLB Related Resources http://www.ed.gov/ nclb/landing.jhtml		Related resources include A-Z Index, NCBL Reauthorization, NCBL Policy, Your State and NCBL, and other tools and models.
OECD – Education http://www.oecd.org/ topic/0,3373,en_2649_ 37455_1_1_1_ 37455,00.html	Organisation for Economic Cooperation and Development	Works to develop and review policies that enhance efficiency and effectiveness of education provisions. Strategies include thematic reviews in specific policy areas, and collecting detailed statistical information on education systems, including measures of the competence levels of individuals. Policies addressed include those implemented within countries for national benefit and those involved in the provision by OECD countries of development aid to build capacity and to spread the benefits of education and training in other countries.
SAT http://www.collegeboard. com/	College Board	SAT Reasoning Test measures critical thinking, reasoning, and writing skills students have developed over time and theoretically the skills they need to be successful academically.

(*continued*)

Table 4.1 (*continued*)

Title and URL	Producer	Content
Science Framework for California Public Schools http://www.cde.ca.gov/re/pn/fd/sci-frame-dwnld.asp	California Department of Education	The Science Framework incorporates and builds on the science content standards adopted by the State Board of Education in 1998. It is the blueprint for reform of the science curriculum, instruction, professional preparation and development, and instructional materials in California, K-12.
Standards and Frameworks http://www.cde.ca.gov/be/st/	California State Board of Education	Information regarding standards and frameworks designed to encourage the highest achievement of every student by defining the knowledge, concepts, and skills that students should acquire at each grade level.
TIMSS http://nces.ed.gov/timss/	U.S. Department of Education, Institute of Education Sciences, National Center for Education Statistics	The Trends in International Mathematics and Science Study (TIMSS) provides reliable and timely data on the mathematics and science achievement of U.S. students compared to that of students in other countries. TIMSS data has been collected in 1995, 1999, and 2003. The United States most recently collected TIMSS data in Spring 2007.
U.S. Department of Education http://www.ed.gov	U.S. Department of Education	Mission is to promote student achievement and preparation for global competitiveness by fostering educational excellence and ensuring equal access. This includes establishing policies on federal financial aid for education, collecting data on America's schools and disseminating research, and focusing national attention on key educational issues.

(*continued*)

Table 4.1 (*continued*)

Title and URL	Producer	Content
US Department of Education, State Index http://www.ed.gov/about/contacts/state/index.html	US Department of Education	Provides access to each state's department of education, state contacts, organizations, accountability, resources, and information.
WestEd http://www.wested.org/	WestEd	A research, development, and service agency, focusing on increasing education and human development within schools, families, and communities. Specific projects include Assessment & Standards Development, Comprehensive School Assistance, Evaluation Research, Health & Human Development, Learning Innovations, Mathematics, Science & Technology, and Teacher Professional Development.

WHY TESTING MATTERS

How does a science teacher select what material to teach in a particular science class, and how to teach it? Although teachers in some senses still have great latitude in selecting materials to present and making homework assignments, national and state standards and curricular objectives are defined for each course and grade level, and legislation also governs some details of what should and should not be taught. The standardized tests that are administered each year are the assessment tool, providing centralized accountability for individual teachers, school districts, states, and the nation, and assuring that classroom instruction does not veer far from the national and state standards. Some tests are administered as part of a worldwide or national review of students' learning of science, other tests are statewide initiatives, and others are elective to students but required for college admission. This array of tests has shifted to the forefront as the factor influencing decisions regarding legislation, funding to school districts, and even affecting property values of homes, as parents are attracted to neighborhoods with a reputation for good schools and science programs.

Science Inquiry and Literacy

In the evolving development of science education based on how students learn, the importance of *inquiry* came into prominence, along with problem-based instruction, really doing science with true unknowns so as to acquire knowledge, technique, and process (Glasgow, 1996). In the era of the Biological Sciences Curriculum Study (1963), Joseph Schwab developed the concept of teaching science as "enquiry," reflecting the concept that scientific principles of enquiry are conceptual structures which can be revised when necessary (DeBoer, 1991, p. 163). The notion of inquiry teaching became so valued that DeBoer (1991, p. 206) argued that *inquiry* is a single word that characterizes the goals of science educators from the 1950s to 1980s. The goal of inquiry became a focal point in science education texts, as with *Teaching Science by Inquiry in the Secondary School* (Sund and Trowbridge, 1967).

As the new progressivism shifted to emphasize science for all and other social issues, a further organizing concept for science education that many people embraced as a central goal is that of science literacy, which includes both the knowledge of science-related concepts and the ability to apply science process skills. These skills involve scientific thinking processes—observing, communicating, comparing, organizing, relating, inferring, and applying (Lowery, 1992), allowing individuals to "think like a scientist" and apply knowledge in authentic contexts. Inquiry-based learning engages participants through a learner-centered approach that emphasizes the development of science process skills, as well as to learn and apply facts, concepts, and the principles of science. Although not profiled, adding other intriguing biological phenomena, with examples relating to children's own pets or other fascinating animals, should excite students and motivate them to learn even more. In contemporary society, individuals must be able to understand and use science (National Research Council, 1996). Understanding the structure and function of living organisms provides a context for health literacy and decision making on behaviors affecting health and medical care that in later life can affect longevity (Baker, Wolf, Feinglass, Thompson, Gazmararian, and Huang, 2007).

Updating his book on the history of science education, DeBoer (2000) provided a refreshing analysis of science literacy and advocated taking a more flexible approach with science content, standards, and testing. He outlined the many laudable goals of science education and argued for a broader, more open-ended approach that would free teachers to innovate.

A Political and Funding "Cauldron"

The controversy historically associated with dissecting and scrutinizing animal bodies is accompanied by a more general disquiet about the entire scientific enterprise. Concerning content, a "cauldron" of differences of opinion regarding curricular offerings on evolution and sex, and whether to include intelligent design, at times threatens to erode the public support and federal funding. The schools seem often to be a focal point for criticism, much of which is directed toward teachers, ultimately making their work harder. Limited class time and inadequate classrooms add challenges to what can be accomplished.

Three areas that are often seen as crucial for biology are potential flash points: evolution, the use of animals both for experiments and dissection in the classroom laboratory, and human reproduction (Rudolph, 2002, pp. 145–146). Rather than respond to external pressures, scientists planning new curricula in the post-Sputnik surge elected to address difficult topics as biologists and not modify the position to serve a minority opinion. The biologists' position held during that period, but in the intervening years, instruction once again has come to be sharply affected by advocacy groups. While this book focuses on uses of animals in dissection and thus relates to the second potential flash point, the other two hot topics add to the challenging environment faced by science teachers. Political controversies concerning specific aspects of content abound. The arguments relating to evolution, creationism, and intelligent design have led to a flood of books and publicity for support or resistance to the teaching of evolution (Campbell and Meyer, 2004; Forrest and Gross, 2004; Shanks and Dawkins, 2004). These political efforts create distractions from consolidating the alignment on standards. Boards of education in Ohio (Holden, 2004) and Texas (Holden, 2003b) are two of many embroiled in controversies. Topics such as sex, sexual activity, sexually transmitted diseases, birth control, and the biology of reproduction are such contentious issues, especially within the United States, that contradictory expectations are set for classroom instruction. Even as educators are seeking to devise a systematic and logical set of learning objectives on human biology that includes reproduction, political groups often put forward legislative or school board policy initiatives that specify very specific details of the content, time frame, or parental permissions that are required for such instruction. Some of these initiatives even effectively muzzle teachers by prohibiting any meaningful discussion of particular topics.

Conflicting Objectives for Science and Health Education

The resounding complexity of methods for setting objectives for public science education in the United States often results in it seeming that schools are buffeted from all directions. The complexity and contradictory objectives arise from several structural factors, all of which contribute to a lower likelihood of allowing teachers to design and equip effective laboratory experiences in biology instruction. The objectives are discussed below:

1. Schools are mostly locally governed. Educational policy is set by local school boards, hopefully in a context consistent with state educational standards and frameworks, and also attentive to national standards and obligated to comply with national requirements such as testing or addressing socioeconomic disparities, so as to obtain federal funding. The development of national and state standards and frameworks is led by educators within a context of the social environment, and may be constrained or contradicted by legislation or policy. This multi-master system of planning educational objectives yields a large number of flash points in the area of science education, where subject matter and social issues are constantly in a state of rapid change and controversy, which vary regionally throughout the country.

2. Biology content appears in the science curriculum, whereas content relating to students' behaviors regarding alcohol, drugs, and sexual activity is assigned to the health curriculum. Expectations of outcomes for biology and health material often differ. Assessment of outcomes for biology is based on learning in tests on science content, but assessment on the health topics is based on students' behaviors: students' self-reports of their engagement in specified behaviors and other criteria such as pregnancy and high school dropout rates.

3. A study of biology provides a scientific framework for understanding the consequences of adverse health practices. While unifying discussions of biology and health would undoubtedly offer a more comprehensive structure for students beginning a lifelong plan for health maintenance, current objectives in health education are more oriented on advocating certain behaviors than on placing behaviors in a context of the human body and physiological consequences.

These three points on educational objectives reflect the entire context of science education and its role in society. They express a variety of societal viewpoints that affect people's opinions affecting education. First, our children's education, science and otherwise, invokes people's most basic hopes, values, and beliefs concerning what is important. It is what they care most about. Secondly, philosophical positions influence the decisions of communities concerning topics such as whether all children should be prepared in their coursework to apply to college. Thirdly, how much individualization should be accommodated in schools is another common question. Viewpoints are affected regarding how children should be grouped in classrooms and whether or not there should be special classes for gifted children or children with disabilities.

Fervent feelings pertain to one's own children, and may also extend to views for society at large. What is our appropriate national role of leadership in the sciences? How will we succeed in continuing as leaders? What steps are appropriate for grooming talented children to enter the scientific workforce?

These questions appear in headlines daily, and are frequently the top-ics of commentaries in the national news media. As one example of a controversial commentary, Charles Murray presented a series of *Wall Street Journal* commentaries characterized as libertarian or right-wing viewpoints on education advocating that fewer people should go to col-lege, even among those scoring in the upper half on IQ tests. Saying, "let us acknowledge the existence of high intellectual ability" (Murray, 2007c), he asserts that people differ in their level of intelligence, and that higher levels of intelligence are associated with higher levels of perfor-mance in classrooms (Murray, 2007a). He argues both for providing spe-cial education for the gifted, and for having a large group of high-IQ people who do not go to college. He advances the notion that an IQ of 115 or higher is necessary for doing well in college, and that no more than 15–25 percent of the population should get a college education. He makes a strong argument for vocational schools and training of craftsmen (Murray, 2007b).

These perspectives contrast with other viewpoints, held with equal fer-vor, that the goal should be toward equalizing performance of all students and furthering diversity at all levels. Jonathan Kozol advocates against segregation and inequality, and speaks on the theme, "Still separate, still unequal" (Stello, 2006). Women and most minority groups are underrep-resented in natural science and engineering degrees compared with the population as a whole (Mervis, 2003). Making gains in raising the per-formance of minority groups is seen as one way to solve the problem of underproduction of scientists in the United States. Despite the scarcity of women entering engineering and the sciences, as compared with the number graduating from college, an emerging equally important issue is the decline in numbers of boys going to college and gaining advanced degrees. There is a striking shortage of highly qualified male applicants to college, indicating some problems in K-12 education for boys, but the source of boys' growing disinterest in education is not well-understood. These political conversations have eclipsed other issues such as concerns with teaching content and technique and dominated much attention for more than a decade, perhaps accounting for the relative silence on topics related to dissection.

Using Funding to Set Policy

The "No Child Left Behind Act of 2001" (NCLB: U.S. Department of Education, 2002b), also known as the Education Sciences Reform Act of 2002, jointly targets academic achievement of disadvantaged students and preparing, training, and recruiting high quality teachers. This legislation added requirements for testing, mandating an increased testing sched-ule, and riveting its importance with funding consequences. In the inter-vening years, education has become a hot topic for front-page news and editorials. NCLB placed a strong emphasis on reading and math, and required that students be provided with qualified teachers. Introducing high stakes into the issue of test results, NCLB linked funding to the performance of schools in the testing program, launching a comprehensive requirement for a program of testing, reporting, and accountability (Wenning, Herdman, Smith, McMahon, and Washington, 2003). Title II grants were dispersed based on the school age population and poverty level in the districts, and then targeted to the schools within the specific districts that had the lowest proportion of highly qualified teachers, had the largest average class size or were identified for needing improvement (National Science Teachers Association, 2002a, b).

The emphasis that historically had been on norm-based testing to monitor the range of performance of students, shifted to criterion-based testing: all students were expected to reach a certain criterion or cut-off point. The program has come under sharp criticism by the National Education Association (2006), for, although NCLB may have laudable goals to improve students' learning, it also has acknowledged problems. One problem concerns the performance of children with learning disabilities, and how their scores should be dealt with; some of them cannot reasonably be expected to reach the required criterion.

One provision of NCLB that hits science in particular is the requirement that teachers be highly qualified. To complete the school scheduling needs, teachers are often necessarily drafted to teach a science course or two that is not in their field of expertise. In California, the legislation requiring qualified teachers has spurred developing a new teaching credential in health education that is served by coursework offered at several of the California State University Campuses (e.g., Chico, Santa Barbara, 2007). The policy statements from the U.S. Department of Education (2004) address challenges in covering science teaching, especially in rural areas, by offering a lower standard of being highly qualified. Achieving compliance by lowering standards is also practiced by states that set their passing scores on state tests at a very easy level compared with the National Assessment of Educational Progress (NAEP) test (King, 2007).

The increase in time that schools spend testing students, and the shift to longer school days, yet sometimes with less math instruction and diminishing time for some other courses, are criticisms in the media of the changes with NCLB (Toppo, 2007). It is argued that since NCLB has placed such a strong emphasis on test scores on reading and math, schools now are aiming lower in their aspirations for science teaching (Janofsky, 2005). Authorities claim that while math and reading are commanding the limelight, science has fallen into a "quiet crisis" (Hovey, Hazelwood, and Svedkawskaite, 2005). Dissension and contention often prevail in discussions of NCLB. Organized opposition in several states has included open rebellion by passing lawsuits or legislation that trumps the federal act (Marks, 2005).

Politics of Standardized Tests

Complicating what used to be the straightforward proceedings of students learning in science classrooms are the consequences of performance on standardized tests. With public funding to school districts increasingly linked to improvements in test results, teachers' efforts must orient on material likely to appear on the tests. Teachers have less latitude to select material simply because it complements lesson plans and will build students' interest in further studies of science.

Test scores are a primary measuring tool when considering diversity and effects of socioeconomic differences among schools. Test results can become the lynchpin for community contention, providing ammunition for arguments concerning property values, school boundaries, busing of students, and assignment of resources. While science test scores are not uniquely singled out as contentious targets, they indicate a likelihood of the United States maintaining reduced scientific and technological leadership in areas ranging from medicine to computers. The extreme leverage swinging from student performance and test scores weighs down what could otherwise be a pleasant learning experience, and disheartens teachers. Knowing that so much emphasis rides on test performance is a development in the current world that most teachers and parents nowadays never encountered during our school years.

MEASUREMENTS OF STUDENTS' SCIENTIFIC PERFORMANCE

Children carry our hopes for the future, both within individual families, and also as a nation. Seeing the United States slip in its standing worldwide on standardized examinations causes alarm that resonates into political action at the local and national level, legislation, shifts in graduation requirements, and changes in funding, a topic we address further in Chapter 6. Test scores absorb valuable classroom time, and the results are significant to families and school districts, directly impacting students more than ever before. A battery of tests is scheduled each year by school districts, and other tests are electively taken by individual students to yield scores to accompany their applications to college. As mentioned, an overview and Web addresses of some key resources pertaining to educational testing can be found in Table 4.1.

Science Test Results: Far-reaching Implications

Poor performance nationally on examinations signals a failure and decline, and an inattention to the younger generation. Property values of neighborhoods shift depending on the perception of performance in the local schools, as expressed in the annual published standardized scores. One complication is that some schools have a lower participation rate in the examination process than others, and the students sitting out exams may not be representative of the school as a whole.

Such concerns reflect a contentious and critical national environment concerning science education and show up almost daily in newspapers with headlines, such as "Is America flunking science?" (Lemonick, 2006). While the United States still grants the most doctorates in science and engineering fields each year, we are 25th worldwide in terms of the percentage of twenty-four-year-olds with science degrees. More and more, we see a value to members of society being conversant with science and having it be near at hand twenty-four hours a day, like water from a tap (Sejnowski, 2003). This may be termed *science literacy*. It seems perhaps we are "Waking up a nation gone drowsy on science" (LaRosa, 2006). One explanation for the poor performance is that we have low expectations for students, expecting only a small number of students to take the advanced math and science courses that are actually essential to intelligent participation in today's world (Barrett, 2005). In a somewhat backwards process, some states that are concerned about achievement rates have considered easing their graduation requirements. Since the report entitled "A Nation at Risk: The Imperative for Educational Reform" (National Commission on Excellence in Education, 1983), attention has focused on the teaching and learning in America's schools compared with those of other advanced nations. An atmosphere of crisis has prevailed since more than two decades ago, with abundant concerns continuing in the published literature, especially regarding science education (Library of Congress, 1990), and this atmosphere continues today. Achieve (2007), a group created by the nation's governors and corporate leaders to help states raise standards has focused on low expectations as a key problem, and directed its energies toward closing the gap from standards and testing to what is expected in the real world of college and the workplace. Achieve reports significant progress among the various states in getting increased alignment between the school expectations and what is required to succeed.

Tests of math performance present an unexpected conundrum, in that the nations where children enjoy math the most show lower performance than nations where children dislike math (Feller, 2006), suggesting that making the subject appealing and palatable may not necessarily be the pathway toward learning excellence.

One survey highlights a finding that for young people, "science is simply not 'cool'" (Vergano, 2006). As mentioned, another aspect to consider is the gender gap in school performance. In general, girls outperform boys, graduating at a higher rate from high school (Mullins, 2006), and then outnumbering boys in higher education. In the professional schools at the University of California in 2004, half of dentistry students were male, with a diminishing proportion ranging from 47 percent to 23 percent in medicine, law, optometry, pharmacy, and veterinary medicine (Gelles, 2006).

Educational crises in the United States have been commonplace in recent decades. As widespread standardized testing has come to the fore as a method of monitoring how students are doing in our nation and local school districts, it often eclipses other concerns that are equally important. Systematic studies and evaluations have made three points that were introduced in Chapter 1.

1. U.S. students are slipping behind worldwide in performance (Trends in International Mathematics and Science Study, 2004) leading to national efforts to raise standards and achievement.
2. Young people from lower economic settings are not attracted to higher education, stimulating close monitoring of diversity and efforts to improve low-performing schools (U.S. Department of Education, 2002a).
3. The quality of current laboratory experiences seems to be poor (U.S. Department of Education, 2002a). Secondary school science laboratories have been sharply criticized for lacking clear goals, not engaging students in discussions, and failing to illustrate how science methodology leads to knowledge, reflecting a lack of focused attention and investment in recent years on this aspect of learning (Singer, Hilton, and Schweingruber, 2005). Despite the concern expressed in these two reports, assessment for these goals does not appear in the required testing programs, nor is the quality of laboratory experience emphasized in the science education standards.

For many students, high school classes and laboratories provide the only formal instruction they receive on their own physiology and health to prepare them as adults to make wise critical decisions concerning their own health management. And, a wide range of constituencies expresses concern regarding the shortage of good teaching resources for the life sciences and the poor quality of hands-on experiences students have with biology materials.

International and National Science Testing

Various measures of education and science performance around the world point to lessening leadership from the United States. Concerning the education level of people ages twenty-five to thirty-five in industrialized nations, the United States ranks ninth in the proportion of people with at least a high school degree, and seventh in the share of those having a college degree (Organization for Economic Cooperation and Development, 2005). Just twenty years ago, the United States ranked first on these two measures. Although the U.S. has many outstanding schools, the variability in quality is a major concern. Additionally, the United States is the second lowest in the proportion of twenty-four-year-olds getting engineering degrees, with only China producing at a lower rate (Davis and Gibbin, 2002).

At a national level, tests are administered by the NAEP, which has administered reading and math tests every two years in the spring since 1969. It offers a kind of benchmark by using the same test for students throughout the United States. Testing in other subjects, including science, is administered in intervening alternate years. In each state, a sample of students is selected to participate in these tests. These test results are published with the title, "Nation's Report Card" (U.S. Department of Education, 2007). They have been invoked and are adapted to provide data for No Child Left Behind (NCLB: United States Department of Education, 2002b). The results of the science testing in 2000 were stable at grades 4 and 8 but showed declines for high school seniors.

The emergence since 2000 of high-stakes testing is a growing trend, as with NCLB, where extra funds, salary bonuses, promotion, or a graduation diploma are tied to a requirement of test achievement (Haury, 2001). Such testing is criticized as killing innovative teaching, commanding excessive time in the curriculum, and being unfair to poor and minority students. Districts struggle to conform to requirements for testing achievement so as to maintain their funding. Since 2002–2003, district applications for Title I grants must include an assurance of participation in the biennial assessments conducted by the NAEP (National Assessment of Educational Progress, 2007).

Although the standards set forth by the United States Department of Education somewhat supersede the states in oversight of education in the areas where testing is required, the states maintain some preeminence in designing their particular testing programs and setting educational goals and curricula. The particular standards for each state are purportedly the basis for that state's testing program and curricula, and states sharply differ in the relative difficulty of the tests they require students to take. This leads to a complex system for textbook review and adoption specific to each state. Textbook publishers need to adapt in particular respects to the priorities of individual states, and in the sciences, some of these state preferences regarding such topics as evolution can become nationally notorious. Even the particular tests used in statewide testing are adapted to some extent for the needs of that particular state.

State Science Testing

As with other forms of testing, requirements for state testing are adopted in an effort to improve performance. Yet, despite efforts in this regard, California, with a comprehensive testing program, had students rank last in the nation in science, as revealed in the "Nation's Report Card: Science 2000" (National Center for Educational Statistics, 2006). As a further complication, California has recently introduced an exit exam as a graduation requirement. This provides an example of how good intentions can go awry. Since this new requirement in 2006, the state's graduation rate fell sharply to 64 percent in 2006 from the previous average over several years of 73 percent (Asimov, 2007a). Passing the California High School Exit Examination is even a requirement for special-education students, as ruled by the California Board of Education. Lawsuits are pending by nonprofit Disability Rights Advocates. Given the stringent pressure on teachers to produce high scores from their classrooms, it should not be surprising that serious allegations are being made of teachers cheating on behalf of their students to achieve a higher yield on scoring (Asimov and Wallack, 2007a, 2007b).

The growing concerns in recent decades regarding the equality of education for different subgroups of the population, and advocacy for diversity, have steadily increased the emphasis on testing. Closer attention is paid to the test performance of ethnic subgroups, socioeconomic microneighborhoods, and geographic areas, as reflected by increased data analyses from all test results in specific years and over time.

Science Testing for College Admission: Coming Back to Dissection

Although not mandated for all students, the SAT tests are typically required for college applications and the results of these tests are watched as closely as the others mentioned above. The SAT tests specifically relate to students going on to higher education, especially those entering the sciences. *Dissection* enters the picture when we consider the question of whether the students are adequately prepared or inspired in their intermediate and high school science to carry them into pursuing science in higher education.

Knowing that we need to focus concern both on science and health literacy for all, and on preparing some students to become outstanding scientists, we should notice the low level of science skills among those students taking the ACT test, usually to prepare for a liberal arts college education. Only 27 percent reached a level in biology likely to earn them a C or higher in college biology (American College Testing: ACT, 2006).

CHAPTER 5

National Versus State Educational Standards: Whither Dissection?

The concepts for students to learn are set forth as learning objectives in published educational standards. Both national standards and those for individual states must be taken into account. Specific guidance for teaching in compliance with the standards sometimes is available in frameworks published at the state level, which are consistent with the standards for the particular state in question. The aggregate of standards and frameworks as presented in national and state documents provide the guideposts that define the subject matter to be addressed at each level in each context. Some key resources pertaining to national science and health educational standards and frameworks, and their Web site addresses, can be found in Table 5.1, along with examples from California, one of the larger states that has extensively developed its standards and frameworks.

These various documents influence the preparation of textbooks and other curricular materials, since states will only adopt textbooks that are viewed as being consistent with the standards and frameworks operative within the state. Thus, teaching materials and lesson plans indirectly follow somewhat from the standards and frameworks, and may need to be adjusted for the specific state. It is important to note that the guiding standards and frameworks do not specify teaching methods, lesson plans, or science laboratories to be used.

In this chapter, we overview some of the most relevant education standards for science and health, especially as they might influence dissection. We describe the national standards for grades 5–8 and 9–12, and also present state standards in California, as an example of how one state addresses these issues in science and health. The National Education Standards described below give strong emphasis to the notion of science literacy

Table 5.1 Educational Standards and Frameworks

Health and Science Standards: California and National

Title and URL	Producer	Content
California State Health Standards http://www.csun.edu/science/standards/health/health_standards.html	California State University, Northridge	California State Content Standards, Health Standards for grades 6 through 12.
California Health Content Standards http://www.cde.ca.gov/ci/he/he/	Health Education, California Department of Education	Information and resources related to health education for students in Kindergarten through grade 12, including the development of health education content standards.
California DOE http://www.cde.ca.gov/	California Department of Education	Mission is to provide leadership, assistance, oversight, and resources so that every Californian has access to an education that meets world-class standards. Goals and objectives include holding local education agencies accountable for student achievement, building local capacity to enable all students to achieve to state standards, professional development in support of teachers, and the effective use of technology.
California DOE Publications http://www.cde.ca.gov/re/pn/fd/		Free downloads of publications available in PDF, including state standards, frameworks, and other department documents.
California Health Framework for Public Schools http://www.cde.ca.gov/re/pn/fd/documents/health-framework-2003.pdf	California Department of Education, California State Board of Education	For Kindergarten through Grade 12, the California State Board of Education and California Department of Education's 2003 edition of the Health Framework for California Public Schools. This edition provides updated information on curriculum development, health literacy, and positive asset development among youths, research-based programs, school safety, and special student populations.

(continued)

Table 5.1 (*continued*)

Health and Science Standards: California and National

Title and URL	Producer	Content
		The framework includes an important change for teachers as it gives greater guidance on what should be taught at each grade level. At the end of Chapter 3, "Health Education," the Grade-Level Emphases Chart suggests specific topics to emphasize in each grade.
National Health Promotion Disease Prevention 5–8 http://www.education-world.com/standards/national/nph/health/5_8.shtml	Education World	U.S. Education Standards, National Standards, Physical Education and Health Standards, Grades 5–8, Grades 9–12
National Health Promotion Disease Prevention 9—12 http://www.education-world.com/standards/national/nph/health/9_12.shtml		
Science Framework for California Public Schools http://www.cde.ca.gov/re/pn/fd/sci-frame-dwnld.asp	California Department of Education	The Science Framework incorporates and builds on the science content standards adopted by the State Board of Education in 1998. It is the blueprint for reform of the science curriculum, instruction, professional preparation and development, and instructional materials in California, K-12.
Standards and Frameworks http://www.cde.ca.gov/be/st/	California State Board of Education	Information regarding standards and frameworks designed to encourage the highest achievement of every student by defining the knowledge, concepts, and skills that students should acquire at each grade level.

(*continued*)

Table 5.1 (*continued*)

Health and Science Standards: California and National

Title and URL	Producer	Content
Supporting Resources		
American Association for the Advancement of Science (AAAS) Project 2061 www.project2061.org	American Association for the Advancement of Science	Long-term AAAS initiative to advance literacy in Science, Mathematics, and Technology.
American Physiological Society (APS) Position Statement on the Use of Animals in Teaching http://www.the-aps.org/publications/tphys/2005html/AugTphys/statement.htm	American Physiological Society	The APS position supports the study of living systems as an essential component of physiology instruction, and believes that teaching laboratories enhance student understanding of physiology, provide experiences that are qualitatively and quantitatively different from those gained through lecture, small group discussion, or multimedia presentations.
Association for Supervision and Curriculum Development (ASCD) http://www.ascd.org/	Association for Supervision and Curriculum Development	A community of educators advocating sound policies and sharing best practices to achieve the success of each learner. Offers broad, multiple perspectives, and addresses aspects of effective teaching and learning, such as professional development, educational leadership, and capacity building.
ASTC Publications http://www.astc.org/pubs/	Association of Science – Technology Centers	ASTC Publications sell books and periodicals for and about the science center field. Topics range from science center management and exhibit planning to research and evaluation.
California Science Teachers Association (CSTA) http://www.cascience.org/	California Science Teachers Association	Represents California science educators in every science discipline at every grade level, Kindergarten through University.

(*continued*)

Table 5.2 (*continued*)

Health and Science Standards: California and National

Title and URL	Producer	Content
		Organizes and participates in statewide reform initiatives and provides opportunities for members who wish to serve on state advisory committees, including framework, standards, and textbook committees.
California Physical Education Health Project http://csmp.ucop.edu/cpehp/standards/hpchallenge.html	California Subject Matter Projects, University of California Office of the President	The California Challenge Standards in Health Education—an effort to foster health literacy and encourage students in the development of skills, knowledge, and attitudes that lead to physically active lifestyles and positive health behaviors. Health literacy is "the capacity of an individual to obtain, interpret, and understand basic health information and services and the competence to use such information and services in ways which are health-enhancing."
California Standardized Testing and Reporting (STAR) Program http://star.cde.ca.gov/star2003/index.asp	California Department of Education	The California Standards Tests in English-language arts, mathematics, science, and history–social science are administered to students in California public schools.
STAR Results http://star.cde.ca.gov/		The 2003 STAR Program results for the state, counties, districts, and schools are available at this site.
Closing the Expectations Gap 2007 http://www.achieve.org/node/844	Achieve, Inc.	This survey updates the efforts of all fifty states to align their high-school standards, graduation requirements, assessments, and accountability systems with the demands of college and work, and finds that at least forty-eight states are now actively engaged in reform efforts of some kind. There is more momentum in the states now than at any time since education reform

(*continued*)

Table 5.1 (*continued*)

Health and Science Standards: California and National

Title and URL	Producer	Content
		became a national priority with the release of *A Nation at Risk* in 1983.
National Association of Biology Teachers (NABT) http://www.nabt.org/	National Association of Biology Teachers	Offers conferences, resources, publications, and workshops.
Position Papers http://www.nabt.org/ sites/S1/index.php? p=26		Position papers include: Administrative Support for Life Science Teachers; Biology Teaching Preparation Standards for Middle & Secondary Teachers; Statement on Teaching Evolution; and, Role of Laboratory and Field Instruction in Biology Education.
National Education Association (NEA) on NCLB http://www.nea.org/ esea/index.html	National Education Association	The NEA offers plans to improve the NCLB Act and discusses ways to influence, fix, and change the law.
National Science Teachers Association (NSTA) http://www.nsta.org/	National Science Teachers Association	Supports standards-based science education and promotes excellence and innovation in science teaching and learning, curriculum and instruction, and assessment. Promotes interest and support for science education. Supports science teachers, science supervisors, administrators, scientists, and others involved in and committed to science education with journals, publications, and conferences.
A Nation at Risk: the Imperative for Educational Reform http://www.ed.gov/ pubs/NatAtRisk/ index.html	The National Commission on Excellence in Education	Report to the Nation and the Secretary of Education, United States Department of Education by The National Commission on Excellence in Education, 1983, includes findings and recommendations.

(*continued*)

Table 5.1 (*continued*)

Health and Science Standards: California and National

Title and URL	Producer	Content
National Research Council (NRC) http://www.nationalacademies.org/nrc/	National Academies	Part of the National Academies, NRC is a private, nonprofit institution that provides science, technology, and health policy advice under a congressional charter. It was organized to associate the broad community of science and technology with the Academies' purposes of further knowledge and advising the federal government, and provides services to the government, the public, and the scientific and engineering communities.
The Nation's Report Card http://nces.ed.gov/nationsreportcard/	National Assessment of Educational Progress	Offers continuing assessment of what America's students know and can do in various subject areas, including math and science.
No Child Left Behind Public Law http://www.ed.gov/policy/elsec/leg/esea02/index.html	U.S. Department of Education	Public Law print of PL 107-110, the No Child Left Behind Act of 2001, Table of Contents, Full-text pdf, and Overview.
NCLB-Related Resources http://www.ed.gov/nclb/landing.jhtml		Related resources include A–Z Index, NCBL Reauthorization, NCBL Policy, Your State and NCBL, and other tools and models.
Organization for Economic Cooperation and Development (OECD)—Education http://www.oecd.org/topic/0,3373,en_2649_37455_1_1_1_1_37455,00.html	Organization for Economic Cooperation and Development	Works to develop and review policies that enhance efficiency and effectiveness of education provisions. Strategies include thematic reviews in specific policy areas, and collecting detailed statistical information on education systems, including measures of the competence levels of individuals. Policies addressed include those implemented within countries for national benefit.

(*continued*)

Table 5.1 (*continued*)

Health and Science Standards: California and National

Title and URL	Producer	Content
U.S. Department of Education http://www.ed.gov	U.S. Department of Education	Mission is to promote student achievement and preparation for global competitiveness by fostering educational excellence and ensuring equal access. This includes establishing policies on federal financial aid for education, collecting data on America's schools and disseminating research, and focusing national attention on key educational issues.
U.S. Department of Education, State Index http://www.ed.gov/ about/contacts/state/ index.html	U.S. Department of Education	Provides access to each state's department of education, state contacts, organizations, accountability, resources, and information.
WestEd http://www.wested. org/	WestEd	A research, development, and service agency focusing on increasing education and human development within schools, families, and communities. Specific projects include Assessment and Standards Development, Comprehensive School Assistance, Evaluation Research, Health and Human Development, Learning Innovations, Mathematics, Science and Technology, and Teacher Professional Development.

that we described in Chapter 4, as highlighted with the most basic Content Standard A, emphasizing "science as inquiry" for grades 5–8.

EDUCATIONAL STANDARDS AND FRAMEWORKS

As detailed in Chapter 3, a swell of support for a new wave of science education reform arose in recent decades; it was launched by two books, *Science for All Americans* and *Benchmarks for Science Literacy* (American Association for the Advancement of Science, 1989, 1993). Widespread prevalence of scientific illiteracy among high school graduates underscored the need for change (Yager, 1996). Profiling the importance of science for all, study of the natural world and presenting a goal of widespread science literacy, these books pointed toward some new approaches that have shaped major reform since then (Ahlgren, 1996).

The intended goals for the content of science education are set forth in the science educational standards, which underwent major revision in 1996. The content states what the students should learn, whereas the curriculum is the way the content is organized and emphasized, including the structure, organization, balance, and presentation of the content in the classroom (National Academies Press, 1996/2007). The key concept of the standards is, *science for all*. While science content is the most obvious aspect of the standards, science teaching, professional development, assessment, science education programs, and science education systems are also important in assuring that all students demonstrate high performance, teachers are empowered, and both teachers and students are supported by educational programs and systems (National Research Council, 1996; Ireton, 1996). To help teachers start moving their teaching toward alignment with the national standards, the National Science Teachers Association (NSTA) developed a project, *Pathways to the Science Standards: Guidelines from Moving the Vision into Practice* (National Science Teachers Association, 1996; Ireton, 1996). Other documents such as frameworks were developed in subsequent years to further assist in this process.

The educational standards serve to set learning goals and objectives that were developed in a broad-based iterative process that represented the National Research Council (NRC), NSTA, and the American Association for the Advancement of Science's Project 2061, and are shared at a national or a state level (Krueger and Sutton, 2001). These goals are necessarily quite general. To flesh out the standards and add specific examples of the concepts to be taught within each standard, frameworks have been developed that enumerate more specifically the particular concepts to be taught within each standard. To create curricula reflecting the science standards, frameworks were initially developed in California, and then similar documents created at a national level, and by some other states. Although they provide more particular detail than the standards, once again the information is still far more general than defining a particular lesson plan and does not suggest the classroom or laboratory methods to use in teaching the concepts. A further problem is the quantity of content in the standards, which often exceeds what is feasible. For example, the volume of material in the California K-8 science standards is considerable, and the specific requirements for health standards can easily exceed the limited class time available.

Stating the goals does not describe the approaches to be used in teaching. How standards are translated into classroom activities at each grade level, and also assuring that the tests are aligned with the standards and teaching activities become a complex issue. Much of this task falls to teachers, who provide the interface between the standards and students, designing activities, selecting materials, and assessing progress (Krueger and Sutton, 2001). This can involve rehearsal of test-taking with preparation of practice tests that are administered each year.

NATIONAL EDUCATION STANDARDS

As mentioned, most children receive no formal instruction in biological science and health or any other science after high school. The general science course in seventh grade is the last such course that is required for all students, and many students do not go on to take the high school course in biology. Some instruction in health is required of students during high school, but the health curriculum is not interwoven with the biology instruction that could provide an understanding of the physiology of disease prevention and avoidance. All adults eventually face myriads of decisions concerning health maintenance. Regarding health, acquiring knowledge and decision-making in adulthood are often challenging, even for those completing a strong biology curriculum in college. Our students who end their formal education after high school have little preparation for the current world of high tech medicine and complex health-care delivery.

The educational goals are defined for each grade level, both nationally and by each state, in published education standards. As noted, the goals for biological science and health are separately addressed, and are implemented in separate courses, generally taught by faculty members with different types of academic preparation. Faculty teaching science in intermediate or high school typically would have earned a teaching credential for teaching science, whereas those teaching health might likely have included some studies of health as part of a physical education credential for intermediate or high school.

Science Education Standards
(Excerpts included from National Academies Press, 1996/2007)

Content Standard A: Science as Inquiry

Dealing with all aspects of science, Content Standard A emphasizes the process and methods of science. Students in grades 5–8 learn about the relationship between explanation and evidence, that scientific inquiry is guided by knowledge, observations, ideas, and questions. Students begin by focusing on questions that can be addressed in scientific investigations. Gathering, analyzing, and interpreting data consolidate the experience of designing and conducting a scientific investigation. Mathematics is an integral part of data presentation. Experience in organizing and presenting data in reports and discussions helps learners practice science and scientific methods of thinking and knowing. Communication of science in writing, drawings, spreadsheets, and computer graphics are part of science education.

The fundamentals of scientific inquiry are central to teaching science. Curricular modules are designed to stimulate inquiry in addressing the standards that are being taught. Examples of instructional materials that are based on the objective of stimulating scientific inquiry are the FOSS (Full Option Science System) modules (Lawrence Hall of Science, 2007) and a recent text (Science Education for Public Understanding Program: SEPUP, 2005).

The National Science Education standards provide a framework for integrating science content with inquiry and personal experience. Emphasizing knowledge of basic science, they embrace a vision of a scientifically literate populace, giving all students an opportunity to learn.

Among the National Science Education content standards for grades 5–8, Standard C, Life Science presents structure and function and is the context where dissection arises. Some of the standards most pertinent to a consideration of dissection are shown below:

Content Standard C: Life Science

1. Structure and Function in Living Systems
    *   Living systems at all levels of organization demonstrate the complementary nature of structure and function. Important levels of organization for structure and function include cells, organs, tissues, organ systems, whole organisms, and ecosystems.
    *   Specialized cells perform specialized functions in multicellular organisms. Groups of specialized cells cooperate to form a tissue, such as a muscle. Different tissues are in turn grouped together to form larger functional units, called organs. Each type of cell, tissue, and organ has a distinct structure and set of functions that serve the organism as a whole.
    *   The human organism has systems for digestion, respiration, reproduction, circulation, excretion, movement, control, and coordination, and for protection from disease. They interact with one another.

. . .

3   Regulation and Behavior
    *   Regulation of an organism's internal environment involves sensing the internal environment and changing physiological activities to keep conditions within the range required to survive (homeostasis).

Content Standard F: Science in Personal and Social Perspectives

1. Personal Health.
    *   Regular exercise. Drug and alcohol abuse. Sexual behavior.

As noted just above, the science standards include within *Content Standard F, Science in Personal and Social Perspectives,* some material that is conventionally addressed in health education. *Science and Technology (Content Standard E)* and *History and Nature of Science (Content Standard G)* are two other content standards, not included here, that bear a strong relationship to consideration of dissection.

Health Education Standards
(Excerpts from American Alliance for Health, Physical Education, Recreation and Dance, 2007; also see University of California Office of the President, 2007a)

Separate from science education, the curriculum for health education provides for the use of knowledge of biological systems of the human body in health promotion, disease prevention, and risk avoidance. Although taught separately and generally lacking in laboratory activities, it is the Health Education Standards that most directly pertain to health maintenance and medical care.

Health Education Standard 1
Students will comprehend concepts related to health promotion and disease prevention to enhance health (only most relevant items are listed below).

Grades 6–8

1. Analyze the relationships between healthy behaviors and personal health.
. . .
4. Describe how family history can impact personal health.
. . .
6. Explain how appropriate health care can promote personal health.
7. Describe the benefits of and barriers to practicing healthy behaviors.
8. Examine the likelihood of injury or illness if engaging in unhealthy behaviors.
9. Examine the potential seriousness of injury or illness if engaging in unhealthy behaviors.

Grades 9–12

1. Predict how healthy behaviors can impact health status.
. . .
4. Analyze how genetics and family history can impact personal health.
5. Propose ways to reduce or prevent injuries and health problems.
6. Analyze the relationship between access to health care and health status.
7. Compare and contrast the benefits of and barriers to practicing a variety of healthy behaviors.
8. Analyze personal susceptibility to injury, illness, or death if engaging in unhealthy behaviors.
9. Analyze the potential severity of injury or illness if engaging in unhealthy behaviors.

STATE EDUCATION STANDARDS

Education in the United States follows a somewhat mixed model between national, state, and local control. We have the National Education Standards which are advisory and implemented depending on the willingness of each state. Each state drafts and adopts its own standards that guide selection of textbooks and preparation of state tests. Textbooks go through review at the state level for adoption, a system that becomes extremely cumbersome for textbook publishers who must comply with the very specific objectives of individual states, such as their treatment of the topic of evolution. Testing is done both at national and state levels, and objectives in testing can reflect either or both the national and state standards.

California Science Education Standards

California is both a very large state and one that pioneered defining its educational standards (while somewhat skirting the national standards), and followed up by setting forth more detailed explanations in educational frameworks. Drawing from California's educational standards, the selected standards presented below relate to the study of animals by using animals or specimens. These illustrate that the national and state standards sometimes overlap and are implemented side by side within the state, in conformity with specific requirements for instruction within that state. Further, the standards for science and health overlap each other.

Grade 7 Science Content Standards
(Excerpted from California State Board of Education, 2007a)

Standard Set 1. Cell Biology

All living organisms are composed of cells, from just one to many trillions, whose details usually are visible only through a microscope. As a basis for understanding this concept, students should know:

a. Cells function similarly in all living organisms.
b. The characteristics that distinguish plant cells from animal cells, including chloroplasts and cell walls.
c. The nucleus is the repository for genetic information in plant and animal cells.
d. Mitochondria liberate energy for the work that cells do and chloroplasts capture sunlight energy for photosynthesis.
e. Cells divide to increase their numbers through a process of mitosis, which results in two daughter cells with identical sets of chromosomes.
f. Students know that as multicellular organisms develop, their cells differentiate.

Standard Set 5. Structure and Function in Living Systems

The anatomy and physiology of plants and animals illustrate the complementary nature of structure and function. As a basis for understanding this concept, students should know:

a. Plants and animals have levels of organization for structure and function, including cells, tissues, organs, organ systems, and the whole organism.
b. Organ systems function because of the contributions of individual organs, tissues, and cells. The failure of any part can affect the entire system.
c. How bones and muscles work together to provide a structural framework for movement.
d. How the reproductive organs of the human female and male generate eggs and sperm and how sexual activity may lead to fertilization and pregnancy.
e. The function of the umbilicus and placenta during pregnancy.
. . .
g How to relate the structures of the eye and ear to their functions.

Standard Set 6. Physical Principles in Living Systems

Physical principles underlie biological structures and functions. As a basis for understanding this concept, students learn:

b. For an object to be seen, light emitted by or scattered from it must be detected by the eye.
. . .
d. How simple lenses are used in a magnifying glass, the eye, a camera, a telescope, and a microscope.
e. White light is a mixture of many wavelengths (colors) and that retinal cells react differently to different wavelengths.
. . .
h. How to compare joints in the body (wrist, shoulder, thigh) with structures used in machines and simple devices (hinge, ball-and-socket, and sliding joints).
i. How levers confer mechanical advantage and how the application of this principle applies to the musculoskeletal system.
j. Contractions of the heart generate blood pressure and that heart valves prevent backflow of blood in the circulatory system.

Standard Set 7. Investigation and Experimentation

Scientific progress is made by asking meaningful questions and conducting careful investigations. As a basis for understanding this concept and addressing the content in the other three strands, students should develop their own questions and perform investigations. Students will:

a. Select and use appropriate tools and technology (including calculators, computers, balances, spring scales, microscopes, and binoculars) to perform tests, collect data, and display data.
b. Use a variety of print and electronic resources (including the World Wide Web) to collect information and evidence as part of a research project.
c. Communicate the logical connection among hypotheses, science concepts, tests conducted, data collected, and conclusions drawn from the scientific evidence.
d. Construct scale models, maps, and appropriately labeled diagrams to communicate scientific knowledge (e.g., motion of earth's plates and cell structure).
e. Communicate the steps and results from an investigation in written reports and oral presentations.

Biology/Life Sciences–Grades 9–12
(Excerpted from California State Board of Education, 2007b) *optional objectives

Standard Set 1. Cell Biology

The fundamental life processes of plants and animals depend on a variety of chemical reactions that occur in specialized areas of the organism's cells. As a basis for understanding this concept, students know:

a. Cells are enclosed within semipermeable membranes that regulate their interaction with their surroundings.
b. Enzymes are proteins that catalyze biochemical reactions without altering the reaction equilibrium and the activities of enzymes depend on the temperature, ionic conditions, and the pH of the surroundings.
c. How prokaryotic cells, eukaryotic cells (including those from plants and animals), and viruses differ in complexity and general structure.
d. The central dogma of molecular biology outlines the flow of information from transcription of ribonucleic acid (RNA) in the nucleus to translation of proteins on ribosomes in the cytoplasm.
e. The role of the endoplasmic reticulum and Golgi apparatus in the secretion of proteins.
. . .
g. The role of the mitochondria in making stored chemical-bond energy available to cells by completing the breakdown of glucose to carbon dioxide.
. . .
*j. How eukaryotic cells are given shape and internal organization by a cytoskeleton or cell wall, or both.

Standard Set 9. Physiology (Homeostasis)

As a result of the coordinated structures and functions of organ systems, the internal environment of the human body remains relatively stable (homeostatic) despite changes in the outside environment. As a basis for understanding this concept, students know:

a. How the complementary activity of major body systems provides cells with oxygen and nutrients and removes toxic waste products such as carbon dioxide.
b. How the nervous system mediates communication between different parts of the body and the body's interactions with the environment.
c. How feedback loops in the nervous and endocrine systems regulate conditions in the body.
d. The functions of the nervous system and the role of neurons in transmitting electrochemical impulses.
e. The roles of sensory neurons, interneurons, and motor neurons in sensation, thought, and response.
*f. The individual functions and sites of secretion of digestive enzymes (amylases, proteases, nucleases, lipases), stomach acid, and bile salts.
*g. The homeostatic role of the kidneys in the removal of nitrogenous wastes and the role of the liver in blood detoxification and glucose balance.
*h. The cellular and molecular basis of muscle contraction, including the roles of actin, myosin, Ca_2, and ATP.
*i. How hormones (including digestive, reproductive, and osmoregulatory) provide internal feedback mechanisms for homeostasis at the cellular level and in whole organisms.

Standard Set 10. Physiology (Infection and Immunity)

Organisms have a variety of mechanisms to combat disease. As a basis for understanding the human immune response, students know:

a. The role of the skin in providing nonspecific defenses against infection.
b. The role of antibodies in the body's response to infection.
c. How vaccination protects an individual from infectious diseases.
d. There are important differences between bacteria and viruses with respect to their requirements for growth and replication, the body's primary defenses against bacterial and viral infections, and effective treatments of these infections.
e. Why an individual with a compromised immune system (for example, a person with AIDS) may be unable to fight off and survive infections by microorganisms that are usually benign.

Standard Set. Investigation and Experimentation. 1

Scientific progress is made by asking meaningful questions and conducting careful investigations. As a basis for understanding this concept and addressing the content in the other four strands, students should develop their own questions and perform investigations. Students will:

a. Select and use appropriate tools and technology (such as computer-linked probes, spreadsheets, and graphing calculators) to perform tests, collect data, analyze relationships, and display data.
b. Identify and communicate sources of unavoidable experimental error.
c. Identify possible reasons for inconsistent results, such as sources of error or uncontrolled conditions.
d. Formulate explanations by using logic and evidence.
. . .
f. Distinguish between hypothesis and theory as scientific terms.
g. Recognize the usefulness and limitations of models and theories as scientific representations of reality.
. . .
j. Recognize the issues of statistical variability and the need for controlled tests.
k. Recognize the cumulative nature of scientific evidence.
l. Analyze situations and solve problems that require combining and applying concepts from more than one area of science.
m. Investigate a science-based societal issue by researching the literature, analyzing data, and communicating the findings. Examples of issues include irradiation of food, cloning of animals by somatic cell nuclear transfer, choice of energy sources, and land- and water-use decisions in California.
n. Know that when an observation does not agree with an accepted scientific theory, the observation is sometimes mistaken or fraudulent (e.g., the Piltdown Man fossil or unidentified flying objects) and that the theory is sometimes wrong (e.g., the Ptolemaic model of the movement of the Sun, Moon, and planets).

California Health Education Standards

Health education, in many states packaged with physical education, may have its own set of standards, separate from either physical education or life science. In addition, most states also have some policies for alcohol, tobacco, and other drug-use education in schools (National School Boards Association, 2007). Along this line, some states, of which Alabama is an example, require a minimum content standard for teaching about these harmful effects and promoting positive healthful behavior. Others provide more specific details, such as California, which designates that two-thirds of Tobacco Use Prevention Education local assistance funds allocated through the California Department of Education are used for education programs on tobacco-use prevention. Health education seems at risk for becoming solely a hodge-podge, and is not paired with the solid instruction in biology that would add credibility to the suggested behavioral restrictions. Currently, the health standards for California are scheduled to be released in 2008, at which time *the law specifically states that schools will not be required to implement them*. Nonetheless, California has a detailed framework that can serve as a guide for health instruction (California State Board of Education, 2003). Outlined below is an example of related work in progress, excerpts from a draft for health standards put forward by the California Physical Education-Health Project (University of California Office of the President, 2007b).

California Challenge Standards in Health Education, California Physical Education-Health Project (University of California Office of the President, 2007b)

Acceptance of Personal Responsibility for Lifelong Health
Expectations:

1. Students will demonstrate ways in which they can enhance and maintain their health and well-being.
2. Students will understand and demonstrate behaviors that prevent disease and speed recovery from illness.

Understanding of the Process of Growth and Development

6. Students will understand the variety of physical, mental, emotional, and social changes that occur throughout life.
7. Students will understand individual differences in growth and development.

BIOLOGY LABORATORIES

Laboratories have long been valued for teaching biology. Yet, in medical physiology, the proportion of physiology courses with animal laboratories has declined from the early 1980s to 2001, from a strong majority of courses, to less than 20 percent (Ra'anan, 2005). Based on a 1983–1984 American Association of Medical Colleges (AAMC) survey reviewed by Ra'anan, the dominating reason for this decline appears to be cost, including the costs of animals, equipment, and supplies, for complying with regulatory oversight, for allocating space for the laboratory, and for the staff salaries. Attitudes of students, faculty, and public were also factors contributing to cutting back on animal use in laboratories. Her analysis sets forth some of the complexity of efforts to compare the use of organisms in laboratories with simpler learning exercises, and reviews some of the informed discussion of this topic.

The American Physiological Society (2007) emphasizes the value of animals in teaching and has pointed to the declining number of schools offering animal labs in connection with physiology and other courses. The controversy about the use of animals, the lack of laboratory space and equipment, the expanding science curriculum, and the higher costs involved in studying animals have contributed to the decline. Their position statement on the use of animals in teaching (American Physiological Society, 2005) points out that the experiences in laboratories enhance students' understanding of physiology as well as their critical-thinking skills, respect for organisms, and appreciation of the complexity of living systems.

A recent comprehensive evaluation of high school laboratories began with a definition: "Laboratory experiences provide opportunities for students to interact directly with the material world (or with data drawn from the material world), using the tools, data collection techniques, models, and theories of science" (Singer, Hilton, and Schweingruber, 2005). Criticisms of laboratories in the report were that they typically lack clear goals, do not engage students in discussion, and fail to illustrate how science methods lead to knowledge. Further problems are that teachers often are not prepared to run science laboratories, state exams do not measure laboratory skills, the quality of equipment shows wide disparities, and educators lack consensus about what laboratory means in the school environment. Among several recommendations to help the experiences achieve the intended learning goals are that the laboratories be: designed with clear learning outcomes in mind; sequenced into classroom science instruction; designed to integrate learning of science content with that about the processes of science; and incorporate ongoing student reflection and discussion. Increasing teachers' capacity to be effective in leading laboratories would require offering teachers effective laboratory experiences and developing more systems of support for them. Other changes were recommended in the organization and structure of high schools. Encompassing details on extensive lists of science topics in state science standards were seen as discouraging teachers from spending much time for effective laboratory learning.

Another recent report on science learning and teaching in prehigh school highlighted that children's thinking is sophisticated beyond the assumptions of current science education, and recommended that frameworks need to take better advantage of children's capabilities (Duschl, Schweingruber, and Shouse, 2007). Focusing on a few core ideas, presenting the process of science and methods, and offering opportunities for interaction and observations were recommended. More substantial teacher development supporting these strategies was strongly recommended.

FUTURE METHODS AND EVOLVING TRENDS

Computer simulations of biological systems are providing new learning opportunities. Typically, assessment instruments show no discernible difference in the learning of students using a computer-derived alternative versus a hands-on animal method. As a rejoinder, however, the point is made that the simulation invariably is less complex than a real organism. The computer simulation may make it easier for the students to focus on specific aspects of the exercise and the difference between the simulation and the real thing may not be evident (Ra'anan, 2005).

Different learning styles are effective for different people. While the field of information technology has not appealed to girls, electronic games and toys have a broader appeal to boys (Balcita, Carver, and Soffa, 2002; Kiesler, Sproull, and Eccles, 1985). The lack of appeal of computers and games to girls is often seen as a problem, but for boys, targeting some materials of this type could be a useful method for reducing the attrition of boys entering the sciences. Video games have educational potential as teaching and learning tools (DeAguilera and Mendiz, 2003). A strong case is being made for bringing the power of games into education by Squire and Jenkins (2003), who use a science-fictional novel by Orson Scott Card as a springboard. In *Ender's Game*, the games teach by encouraging experimentation, exploration, and innovation in a setting that the authors describe as a constructivist utopia.

Games have not been effectively harnessed in science laboratories. Many students play educational games and are highly engaged in them, but they fail to actively reason and thus do not learn from the game. One approach is to monitor students' emotional reactions as well as their cognitive states while interacting with the game (Conati and Zhou, 2002). By modeling the students' cognitive and emotional states as they interact with games, much more can be done to develop interventions to trigger their constructive reasoning (Conati and Klawe, 2002).

The presentation of standards has become a unified effort that is well known and supported by educators. Science literacy assumes a primacy with a realization of its importance for everyone, not just for would-be scientists (Raloff, 2001). Middle schools have been a focus of alternative curricular materials that often are presented in modules, and emphasize experimentation rather than being based on textbooks. Project 2061, an educational effort of the American Association for the Advancement of Science (2005), steadily emphasizes "science literacy for a changing future" and has targeted improved K-12 science materials, and alignment of assessment to standards. Yet, teachers tend to teach the way they were taught and use didactic lecturing as they were taught. Thus, they may avoid inquiry-based methods because it is exceedingly difficult for them to incorporate such teaching methods. As a preparation for teachers, it has been pointed out that the teaching enterprise in the university needs to use active learning, problem-based learning, and inquiry-based learning in courses for teachers (Bonner, 2004).

Some of the most exciting developments in science education are occurring at science museums such as the Marian Koshland Science Museum of the National Academy of Sciences in Washington, D.C. (Smith, 2004). Targeted to appeal to anyone from age thirteen and up, this museum seeks to stimulate discussion among families. The high school students staffing the intern program on weekends help run the hands-on science activities on weekends and make them more accessible to students.

Without a doubt, dissection of animal cadavers is on its way out, if for no other reason than the cost. However, the advent of computer imaging technology, supplemented with plastinated models or a few prosected animal specimens, now provides the medium to present the material typically covered in dissection in an inexpensive and educationally effective manner. Although this avenue is just beginning, the most notable example to date is the *Visible Human*, available through the National Library of Medicine (2007), offering high-resolution views of the human body.

CHAPTER 6

Legislation and Regulations Related to Using Animals and Dissection in Teaching

In Chapter 5, we have seen that a complex array of educational guidelines and standards guide the teaching of science, as it does other disciplines. In this chapter, we see that superimposed on the standards are a set of laws that dictate what teachers must or cannot teach or how the teaching should be done. This especially applies to courses where dissection is involved. Education in the United States is commonly regarded as being under the oversight of each state, but many national laws set specific requirements for federal funding that preempt the states' decision-making. Then, within the national and state oversight laws, a tradition of local governance of schools exists for the local districts. Thus, we have a situation where in all matters a complex of laws, regulations, and policies from the national, state, and local level set the stage for what is supposed to occur in the classroom with regard to science content to be taught, guidelines for the funding for science education, requirements for student participation in dissection or not, and regulation of the welfare of animals in classrooms. This chapter addresses the complex legal and regulatory aspects of the situations in science education, especially where animals are used. This is a challenging and historically contentious topic that reflects deeply held values. Resources pertaining to the four legislative areas are as follows: Legislation and Policy on Biology and Health Instruction (Table 6.1a), Legislative Impacts on Funding for Science Education (Table 6.1b), Legislation regarding Students' Participation in Dissection (Table 6.1c), and Legislation and Guidelines Regarding Animal Welfare and Use of Animals or Specimens in Classrooms (Table 6.1d).

Table 6.1 Laws and Regulations

a. Legislation and Policy on Biology and Health Instruction

Title and URL	Producer	Scope
Health Framework for California Public Schools http://www.cde.ca.gov/re/pn/fd/ts/health-framework-2003.pdf	California Department of Education, California State Board of Education	For Kindergarten through Grade 12, the California State Board of Education and California Department of Education's 2003 edition of the Health Framework for California Public Schools. This edition provides updated information on curriculum development, health literacy, and positive asset development among youths, research-based programs, school safety, and special student populations.
		The framework includes an important change for teachers as it gives greater guidance on what should be taught at each grade level. At the end of Chapter 3, "Health Education," the Grade-Level Emphases Chart suggests specific topics to emphasize in each grade.
Health Promotion Disease Prevention 5-8 http://www.education-world.com/standards/national/nph/health/5_8.shtml	Education World	U.S. Education Standards, National Standards, Physical Education and Health Standards, Grades 5–8, Grades 9–12
Health Promotion Disease Prevention 9-12 http://www.education-world.com/standards/national/nph/health/9_12.shtml		
Health Research Extension Act of 1985 http://grants.nih.gov/grants/olaw/references/hrea1985.htm	National Institutes of Health, Office of Extramural Research	Health Research Extension Act of 1985; Public Law 99-158, November 20, 1985; "Animals in Research" DHHS. National Institutes of Health.

(*continued*)

Table 6.1 (*continued*)

a. Legislation and Policy on Biology and Health Instruction

Title and URL	Producer	Scope
Humane Education Laws http://www.teachhumane. org/laws.html	Humane Education Advocates Reaching Teachers (HEART)	Working to help schools comply with Humane Education laws currently in existence in New York State, the HEART site also provides links to humane education laws in several other states.
National Health Education Standards http://www.aahperd.org/ aahe/pdf_files/standards. pdf	American Alliance for Health, Physical Education, Recreation, and Dance	Information is cited from the prepublication document of National Health Education Standards, pre-K-12, American Cancer Society.
State-level School Health Policies http://www.nasbe.org/ HealthySchools/States/ Topics.asp&?Category= A&Topic=7	National Association of State Boards of Education (NASBE)	State-by-State Alcohol, Tobacco, and Drug Use Education.

THE CONTENT: INSTRUCTION IN BIOLOGY OR HEALTH

Anyone wishing to be literal in interpreting the full body of educational standards and frameworks, along with the laws, regulations, and policies relating to science and health education, will quickly be confused and embroiled in the "cauldron" of conflicting requirements. As one example, fulfilling the instructional requirements in the educational standards, as described in Chapter 5 and put forward for teaching about the immune system, almost inevitably violates other requirements in some states or districts for avoiding talking about the organs of reproduction. Setting an artificial boundary between biology and health, and effectively muzzling teachers from teaching about the reproductive system as part of biology, are examples of us being shortsighted in preparing students for their lifetimes when they and their family members will be dealing with diseases in which they would benefit from some understanding of their bodies and physiology. Opportunities are missed for offering knowledge for effective prevention of diseases. Dissection offers a way around this dilemma because dissection places the entire organism in view without required erasure of certain organ systems, and this whole-bodied realism may be a part of dissection's continuing appeal for teachers who are presented conflicting teaching objectives.

In the current climate, the program mandate, "No Child Left Behind," places the emphasis on mathematics and English, relegating science education to the background. Nonetheless, required national and state educational standards are in place for instruction in biology and health at various grade levels. National regulation has gained weight by hinging major funding upon the test results of school districts. These requirements may impact subgroups of students differentially and relate to problems in the broad context of public school education. For example, many authorities feel that children in special education classes may not thrive when in mainstreamed classrooms or even in public schools, which can lead to contentious conflicts and litigation between the parents involved and the particular districts. Recent judicial rulings allow such parents to represent themselves in court when suing school districts (Asimov, 2007b; Biskupic, 2007). Spanish speaking families have sought in California to require that their children be examined in Spanish rather than English, but this lawsuit was recently dismissed (*California Aggie*, 2007). A California requirement for students to pass an exit test before high school graduation has precluded many students from graduating (with a consequent lowered graduation rate), even though by previous requirements they would have had a diploma (*San Francisco Chronicle*, 2007).

Most curricular objectives for biology and health evolve as educational standards from coordinated efforts of educators in national and state departments of education. However, this collaborative, well-thought-out approach is preempted by political intervention that also plays a role in setting educational objectives. The topic of evolution versus intelligent design is perhaps the most striking and well-known example of biology instruction being influenced by political forces. In such cases, legislation is created by a state that sets specific requirements for instruction on particular topics. Another example is instruction regarding risky sexual behavior or using harmful substances. For many years, California, like many other states, has had some of these requirements on alcohol and drug use, and instruction is offered in health classes on these topics. Instruction on reproduction is also mandated in California.

An underlying theme of these mandated teaching objectives is that they are oriented on specifying a particular type of behavior by students and the goal of instruction is to alter the students' behavior, rather than stimulating learning of particular content in a general area of knowledge. The standards for health, perhaps for this reason, seem less content oriented than the science standards, and instead express the expectation of altering human behavior, a more challenging goal than increasing learning.

Use of Alcohol, Tobacco, or Other Drugs

Most states have passed specific legislation pertaining to required instruction for students in secondary education pertaining to the use of alcohol, tobacco, and drugs (National Association of State Boards of Education, 2007). This subject matter content typically appears within the health standards rather than the science standards and is likely to be taught by teachers who majored in physical education and minored in science. This required instruction on the use of alcohol and other potentially harmful substances has been given in California for over half a century, often interposed with classes in driver education. Being isolated from the study of biology, this instruction is likely to be less instructive than a curriculum considering alcohol's effects on the various body systems.

Sexual Activity

Instruction on sexual behavior and potential exposure to sexually transmitted diseases often becomes an emotionally and politically charged issue in communities with some groups advocating for the instruction of the material, and others wanting to preclude it. As with presentation of information on alcohol and tobacco, the instruction would be more effective and long-lasting if included within a context of learning biology, dealing with the anatomy and physiology of the reproductive system and disease processes thereof.

As we noted in Chapter 5, the national science standards specify teaching in grades 5–8 about the reproductive system under *Content Standard C: Life Science. Structure and Function in Living Systems* and also about sexual behavior under *Content Standard F: Science in Personal and Social Perspectives. Personal health.* In health education standards, sexual behavior is addressed in a context of disease prevention and healthy behaviors, emphasizing advocacy for certain healthy behaviors. However, as we discussed earlier, there is a complicated interplay among standards and policies in which state standards and local-school board policies trump the national science and health standards.

Instruction concerning sexual behavior, the reproductive system, birth control methods, sexually transmitted diseases, the immune system, and pregnancy are subjects that are addressed in exquisite detail in the media and make their way into many policies of school boards and even state policy initiatives, some of which conflict with the standards described in the previous paragraph. Many states and local school boards seek to micromanage the content presented to students on these specific topics, including regulating the extent to which parents need to request or decline this instruction for their children, and setting the maximum amount of class time to be devoted to the material.

Advocacy for abstinence prohibiting mention of birth control is a significant national movement with legislative initiatives in several states. Specialized curricular materials are available for purchase for this purpose and are used in many states, e.g., Choosing the Best Path: A Curriculum for 7th Grade (Choosing the Best, 2007), WAIT Training (2007), and A.C.

Green's Game Plan Abstinence Program (Project Reality, 2001). This is an active area for legislation at the state and local level and is regularly monitored by SIECUS (Sexuality Information and Education Council of the United States, 2005), NARAL Pro-Choice America (National Association for the Repeal of Abortion Laws, 2007), and the Kaiser Network (Kaiser Family Foundation, 2007) that provide summary overviews of legislative and regulatory developments. In different states, legislation has been passed both requiring and prohibiting abstinence-only education. Thus, a teacher moving from one state to another could face requirements that are exactly opposite.

School board policies also can become very specific in setting requirements for the presentation of information on sexual behavior and reproduction. Two examples of administrative procedure documents are from the San Diego Unified School District (2005) and the Torrance Unified School District (2005) requiring parental notification for any instruction on human reproductive organs. In the state of Arizona, the relevant statutes and Board of Education rule pertaining to instruction on sex education and acquired immune deficiency syndrome (AIDS) are included in the document of the Comprehensive Health Education Standards (Arizona Department of Education, 1997). Conditions for offering sex education through elementary grades in Arizona have specified that this elective instruction is only taken at the written request of the parent. Boys and girls are to be separated and no homework or tests given. At the high-school level, instructional materials for the unit are reviewed, and the emphasis is to be on the power of individuals to control their own personal behavior.

Given the overwhelming regulatory efforts pointing at education on reproduction, it should not be surprising if teachers completely duck addressing the national science standards pertaining to the reproductive system and sex behavior and simply touch on the national health standards specifying healthy sexual behaviors. In many teaching situations, teachers effectively are prevented from presenting meaningful and informative material that would be useful in the area of reproduction.

"NO CHILD LEFT BEHIND" AND FUNDING
FOR SCIENCE EDUCATION

A strict pattern of accountability to the national Secretary of Education is required for the testing entity, National Assessment of Educational Progress (NAEP), and for any state that wishes to receive a Title I federal grant, or for any local education agency that receives such a subgrant. *No Child Left Behind* sets in place a cascade of legislated requirements, including mandatory biennial, national, and state testing in reading and mathematics in grades 4 and 8. If time and money allow, assessments also may be conducted in additional subjects in grades 4, 8, and 12. A corollary of these requirements is that state participation in testing assessments

Table 6.1 Laws and Regulations

b. Legislative Impacts on Funding for Science Education

Title and URL	Producer	Scope
NCLB Act http://www.ed.gov/policy/elsec/leg/esea02/index.html	U.S. Department of Education	No Child Left Behind Act, Table of Contents, Full Text.
http://www.ed.gov/nclb/landing.jhtml		U.S. Department of Education NCLB Act Overview.
National Education Association (NEA) view of NCLB Act http://www.nea.org/esea/index.html	National Education Association	NEA perspective and positive agenda for the No Child Left Behind Act.
National Science Teachers Association (NSTA) Legislative News http://www3.nsta.org/main/news/stories/nsta_story.php?news_story_ID=47257	National Science Teachers Association	Professional development opportunities for science educators and Improving Teacher Quality State Grants information.

for other subjects is voluntary, e.g., writing, science, history, geography, civics, economics, foreign language, and arts. The state plan that is submitted to the national Secretary of Education needs to include an assurance of participation in biennial state-level NAEP testing in reading and mathematics at grades 4 and 8. These requirements mean that school district funding overall is dependant on the local students' performance in reading and mathematics, with science education relegated to the back burner.

The requirement of *No Child Left Behind* that teachers be qualified is complicated by the muddled mixture of science and health education. What requirements are adequate for teaching science or health? The new teaching credential being offered in California in health offers qualification for teaching health. Whether this is raising the bar above the physical education teaching credential or lowering the bar below a science teaching credential is a matter of debate. Long before the *No Child Left Behind* legislation, Juliana Texley (1992) raised an alarm about ill-prepared teachers who acquired an endorsement for teaching health education by taking a few zoology classes. She cited a survey by the National Science Teachers Association (NSTA) that reported over half of life science classes in the United States were being taught by nonmajors in science. Hence, they had little knowledge of laboratory work in genetics, biochemistry, or microbiology, but were familiar with dissection.

OPTING OUT OF DISSECTION

The effect of legal intervention discussed above regarding teaching about human reproduction is paralleled by comparable legislation with regard to dissection. The emotional dilemmas for some students associated with dissecting animals has led to legal action and legislation, starting with the California suit filed and won by Jennifer Graham in 1986 after her grade was lowered for refusing to participate in dissecting a frog. Subsequent legislation in several states now offers students the option of declining to dissect. Here is an example where legal intervention from outside led to legislative reform.

Opposition to using animals in science courses gathered momentum as a result of several cases where students filed suit or refused to complete coursework where use of animals was required, leading to a media storm on the topic during the late 1980s (Holden, 1987; McCabe, 1990). The California Assembly introduced a bill in 1987 to authorize students at all levels of instruction in all public and private schools, and from K-12 through the university level the option to refuse to participate in any course requirement involving the harmful or destructive use of animals. The bill applied even to dead animals or animal parts. Teachers were to work with such students to develop and agree upon an alternative education project. Scientists at Stanford University were among those opposing this version of the bill (Hamm and Blum, 1992). Though not enacted, this bill reflected the growing concern regarding requiring students to participate in laboratories involving animals. The amended version that was passed applies only to public school grades 1–12, and allows teachers to either have the pupil pursue an alternative educational project or be excused from the project.

Most of the legislative effort concerning precollege uses of animals has concerned students being able to decline participating in dissection. Fifteen states have passed legislation or resolutions requiring that students have options other than dissecting if they choose not to participate therein. Florida, California, Pennsylvania, New York, Rhode Island, Illinois, Virginia, Maine, and Louisiana have laws upholding a student's right to choose alternatives to dissection without being penalized (Humane Society of the United States, 2007d; National Association for Biomedical Research, 2007). Resolutions by the state's department of education have been passed in Maine, Maryland, Massachusetts, and New Mexico.

Organizations supporting this legislation offer counseling support for students who find themselves in difficult positions regarding dissection. Web sites and consultation are available to lend further support of this type to the students (American Anti-Vivisection Society, 2007; Animalearn,

Table 6.1 Laws and Regulations

c. Legislation Regarding Students' Participation in Dissection

Title and URL	Producer	Scope
The Dissection Hotline http://www.navs.org/site/ PageServer?pagename=ain_ edu_dissection_hotline	National Anti-vivisection Society (NAVS)	The NAVS Dissection provides information, counseling, and support for students, parents and teachers who object to dissection. Any student, teacher, parent, or school district interested in obtaining information on alternatives to dissection in the classroom is encouraged to call for help.
International Dissection Laws and Policies http://www.neavs.org/esec/ legislation/legis_ INTERNATIONAL_ dissection_laws.htm Legislation Index http://www.neavs.org/esec/ legislation/legislation_index. htm Summary of U.S. Dissection Laws http://www.neavs.org/esec/ legislation/legis_US_ dissection_laws.htm	New England Anti-Vivisection Society, NEAVS' Ethical Science & Education Coalition (ESEC)	International Dissection Laws and Policies; Table and Index summarizing U.S. Dissection Laws.
Overview of Laws and Regulations in U.S. Society http://www.osera.org/ drftcatET1EIXPH.html	Ohio Scientific Education & Research Association (OSERA)	Federal laws and regulations governing the use of animals in scientific inquiry.
State Legislation Dissection Laws http://www.hsus.org/ animals_in_research/animals_ in_education/dissection_laws. html	Humane Society of the United States (HSUS)	Student Choice Laws are choice-in-dissection laws or policies allowing students to refuse to participate in classroom dissections. These laws apply to kindergarten through high school.
Student Choice Laws http://www.navs.org/site/ PageServer?pagename=ain_ edu_student_choice_laws	National Anti-vivisection Society	Student choice laws, organized by state.

(continued)

Table 6.1 (*continued*)

c. Legislation Regarding Students' Participation in Dissection

Title and URL	Producer	Scope
Student Choice Laws http://www.navs.org/site/ Search?query=Student+ Choice+Laws		
State Laws and Legislation http://animalearn.org/ lawsandlegislation.php	Animalearn	Dissection laws and legislation, organized by state.

2007; Association of Veterinarians for Animal Rights, 2007; Humane Society of the United States, 2007b; Physicians Committee for Responsible Medicine, 2007).

USE OF ANIMALS AND SPECIMENS

In a study in Stanislaus County, California, a majority of teachers kept classroom pets, common examples being fish, reptiles (iguanas, snakes, turtles, and skinks), and small mammals such as hamsters, guinea pigs, gerbils, mice, and rats (Zasloff, Hart, and DeArmond, 1999). Some (35%) had more than one type of animal. Nonliving specimens from animals, such as feathers, bones, and nests were used by nearly three-quarters of the teachers. Teachers mentioned drawbacks of keeping classroom pets, including the responsibility and cost of ongoing care, potential health and safety issues for children, and concerns about the welfare of the animals. They believed the animals were popular and effective foci of instruction, especially through the primary grades.

Concerning the possible health issues for children, reptiles in classrooms raise a risk of salmonellosis, which is especially dangerous for anyone who is immunocompromised. Exposure to formalin-preserved specimens is also problematic. Anyone wearing contact lenses needs to be cautious, and the fumes cause respiratory discomfort to anyone lacking adequate ventilation in the presence of the formalin.

No federal legal requirements or oversight pertain to use of animals in the lower grades. The federal government has established a system for overseeing other uses of live animals, as expressed in the Animal Welfare Act and the *Guide for the Care and Use of Laboratory Animals* (U.S. Department of Agriculture, 1966/2002; Institute of Laboratory Animal Research, 1996). With the notable exception of precollege instruction, use of animals for research, teaching, and testing are subjected to a detailed protocol review process involving a local institutional animal care and use committee

Table 6.1 Laws and Regulations

d. Legislation and Guidelines Regarding Animal Welfare and Use of Animals or Specimens in Classrooms

Precollege

Animal Protection Institute (API) Legislation http://www.api4animals. org/b4_legislation.php	Animal Protection Institute	Organizes and updates regularly information on both federal and state legislation.
Animal welfare legislation, regulations and guidelines, 1990–1995 http://www.nal.usda.gov/ awic/pubs/oldbib/qb9518. htm	Tim Allen, Animal Welfare Information Center, National Agricultural Library	This publication includes citations related to codes of practice on animal welfare and international regulations.
California Animal Laws Handbook http://www. californiastatehumane. org/catalog.htm	State Humane Association of California, Pacific Grove, CA	A compilation of laws relating to animal welfare and control in the state of California, 2007 edition.
Humane Ed Laws http://www.teachhumane. org/laws.html	Humane Education Advocates Reaching Teachers (HEART)	Working to help schools comply with Humane Education laws currently in existence in New York State, the HEART site also provides links to humane education laws in several other states.
Humane Education Programs http://www.humaneedu. com/humeduprog.html	Humane Education Programs	An animal protection advocacy group, their site offers lesson plans and informational references on humane education, with the goal of helping teachers incorporate humane education into their curriculum.
Institute for Laboratory Animal Research (ILAR) http://dels.nas.edu/ilar _n/ilarhome/	Institute for Laboratory Animal Research	ILAR prepares authoritative reports on subjects of importance to the animal care and use in community, as well as develops and makes available scientific and technical information on laboratory animals and other biological research resources.

(continued)

Table 6.1 (*continued*)

---

d. Legislation and Guidelines Regarding Animal Welfare and Use of Animals or Specimens in Classrooms

---

For educators and students
http://dels.nas.edu/ilar_n/ilarhome/educators.shtml

ILAR Principles and guidelines for the use of animals in precollege education
http://dels.nas.edu/ilar_n/ilarhome/Principles_and_Guidelines.pdf

Intel International Science and Engineering Fair (ISEF) http://www.sciserv.org/isef/	Science Service	Since 1921, Science Service's mission has been to advance public understanding and appreciation of science among people of all ages through publications and educational programs. Intel is the title sponsor of the ISEF, which is the world's largest precollege science fair, bringing together nearly 1,500 students from more than 40 nations to compete for scholarships, tuition grants, internships, and scientific field trips.
National Association for Biomedical Research (NABR) Animal Law http://www.nabr.org/AnimalLaw/Education/Dissection.htm	National Association for Biomedical Research	The Dissection in Education section provides access to valuable resources related to animal law advocacy and federal laws, policies, and regulations.
National Association of Biology Teachers (NABT) http://www.nabt.org/ NABT Position Statements http://www.nabt.org/sites/S1/index.php?p=26	National Association of Biology Teachers	NABT is the primary professional association for those in life science education. Resources and materials useful to teaching include Biology Teaching Preparation Standards for Middle & Secondary Teachers, Ethics Statement for Biology Teachers, NABT's Statement on Teaching Evolution, The Role of Biology Education in Addressing HIV &

(*continued*)

Table 6.1 (*continued*)

---

d. Legislation and Guidelines Regarding Animal Welfare and Use of Animals or Specimens in Classrooms

---

		AIDS, Role of Laboratory and Field Instruction in Biology Education, and The Use of Animals in Biology Education.
National Science Teachers Association (NSTA) http://www.nsta.org/	National Science Teachers Association	NSTA works to provide and expand professional development to support standards-based science education, while also supporting innovation in science teaching and learning, curriculum and instruction development, and improved assessment.
Official Positions http://www.nsta.org/ about/positions.aspx		Positions include Responsible Use of Live Animals and Dissection in the Science Classroom; Science Competition; Elementary School Science; and National Science Education Standards.
Intel Science Talent Search (STS) http://www.sciserv.org/ Sts/	Science Service	Since 1942, Science Service has organized this precollege science contest for high school seniors. Intel is currently the sponsor of this nationwide competition.

College

2007 Report of the AVMA Panel on Euthanasia http://www.avma.org/ issues/animal_welfare/ euthanasia.pdf	American Veterinary Medical Association, Panel on Euthanasia	Accepted methods of euthanasia of animals in research and animal care and control facilities.
Association for Assessment and Accreditation of Laboratory Animal Care, International (AAALAC) http://www.aaalac.org/	Association for Assessment and Accreditation of Laboratory Animal Care, International	AAALAC International is a private, nonprofit organization that promotes the humane treatment of animals in science through voluntary accreditation and assessment programs. The accreditation process requires research programs to demonstrate that they go beyond the minimum standards required by law in order to achieve excellence in animal care and use.

(*continued*)

Table 6.1 (*continued*)

---

d. Legislation and Guidelines Regarding Animal Welfare and Use of Animals or Specimens in Classrooms

---

Animal Protection Institute (API) Legislation http://www.api4animals.org/b4_legislation.php	Animal Protection Institute	Organizes and updates regularly information on both federal and state legislation.
Animal Care Publications, Animal Welfare Act, Regulations, and Standards http://www.aphis.usda.gov/ac/publications.html	Animal Care, USDA	AWA Regulations; Animal Welfare Act Summary; 2000 Report on the AVMA Panel on Euthanasia; Fact Sheets; animal Care Manuals; Animal Care Reports; Animal Care Violation Summaries.
Animal Welfare Act and Legislation http://www.nal.usda.gov/awic/legislat/usdaleg1.htm	Animal Welfare Information Center, U.S. Department of Agriculture,Agricultural Research Service National Agricultural Library	Questions and answers about the Animal Welfare Act and its regulations for biomedical research institutions; Animal Welfare Act interpretive summaries; Animal Welfare Act as amended; Code of Federal Regulations, Title 9, Chapter 1, Subchapter A—Animal Welfare.
Code Federal Regulations Title 9: Animals and Animal Products http://www.access.gpo.gov/nara/cfr/waisidx_06/9cfrv1_06.html	U.S. Government Printing Office	Title 9—Animals and Animal Products (this index contains parts 1–199) Chapter 1: Animal and Plant Health Inspection Service, Department of Agriculture.
Guide for the Care and Use of Laboratory Animals http://www.nap.edu/readingroom/books/labrats/index.html	National Academy of Sciences	The Guide (1996) is characterized as a living document, subject to modification with changing conditions and new information. The emphasis on performance goals as opposed to engineering approaches, the purpose being greater enhancement of animal well-being.
Public Health Service (PHS) Policy on Humane Care and Use of Laboratory Animals http://grants.nih.gov/grants/olaw/references/phspol.htm	National Institutes of Health, Office of Extramural Research	This 2002 reprint of the PHS Policy on Humane Care and Use of Laboratory Animals includes amendments regarding PHS Animal Welfare.

(*continued*)

Table 6.1 (*continued*)

d. Legislation and Guidelines Regarding Animal Welfare and Use of Animals or Specimens in Classrooms

Society for Protective Animal Legislation (SPAL) http://www.saplonline.org/	Animal Welfare Institute (AWI)	Specifically dedicated to the passage and promotion of federal, international, and local legislation to ensure and protect animal welfare, SPAL is a good resource for locating current legislation and guidelines.
Transportation,Sale, and Handling of Certain Animals http://www.aphis.usda.gov/ac/publications/AWA/AWAindex.html	Animal Care, U.S. Department of Agriculture (USDA)	Animal Welfare Act United States Code Title 7—Agriculture Chapter 54—Transportation, Sale, and Handling of Certain Animals.
U.S. Department of Agriculture (USDA) Animal and Plant Health Inspection Service (APHIS) Policy Manual http://www.aphis.usda.gov/ac/polmanpdf.html	USDA, APHISAnimal Care	USDA, APHISAnimal Care Policy Manual.

which gives feedback prior to approving the proposed work. However, a further exception is that rats, mice, birds, fish, and cold-blooded animals are exempted by the Animal Welfare Act, but are covered by the more comprehensive Public Health Service requirements set forth in the *Guide for the Care and Use of Laboratory Animals*, for which AAALAS accreditation (Association for Assessment and Accreditation of Laboratory Animal Care International, 2007) affirms compliance. Complying with these latter requirements is required for institutions receiving federal funding, which would include virtually all colleges and universities.

The laws passed pertaining to laboratory animals in higher education typically focus on inspection of animal facilities and control of pain and distress (Orlans, 1991, 1993, 1998). Painful and distressing procedures are specifically clarified in the USDA Policy 11 (U.S. Department of Agriculture, 1997). We note also that with few exceptions, use of animals for education typically receives much less attention than use in research, where advocates emphasize seeking alternatives (Stokes and Jensen, 1995; Stokes, 1997). Use of animals in higher education for teaching accounts for relatively few animals in comparison with the other uses, mainly research.

Legislation regarding animals is oriented on preventing pain and distress of animals. The specific provisions deal with reducing the likelihood, and extent and number of occasions when animals experience pain or distress. The death of animals per se is not targeted in legislation, except to assure that the animal does not experience pain and distress while dying.

Animal Welfare Legislation Excludes Precollege Schools

Some informal voluntary guidelines exist for precollege schools, but they are relatively imprecise and do not address many problems that arise. Historically, the inhumane treatment of animals in some science fair projects revealed difficulties in not providing structured supervisory guidelines to ensure humane treatment of animals.

The lack of oversight or federal guidelines for the use of animals at the precollege level leaves a vacuum, relying on schools to provide appropriate structure, judgment, and guidance for use of animals that arise. The wide range of animal uses in classrooms includes every type of classroom pet, animals used in science projects, and dissection. Informal science education, such as 4-H, is another setting where children have involvement with animals. One possible concern is that their group leaders may have little formal preparation for supervising these projects.

Optional Informal Guidelines for Precollege Schools

A growing focus on some of the inappropriate uses of animals in secondary schools, often at science fairs, has led to preparing voluntary guidelines as a reference for precollege educators. Most widely disseminated of these is the document prepared by the Institute of Laboratory Animal Resources (2006): "Principles and Guidelines for the Use of Animals in Precollege Education," that includes ten principles. These prescribe the use and care of animals, but are not legally binding and include no oversight to assure proper animal care.

The National Science Teachers Association (NSTA: 2005) has adopted a position statement that offers very specific recommendations for teachers pertaining to the use of live animals in the classroom and also for dissection. This statement is a forthright presentation of a topic that is primarily ignored in the published literature by science teachers. The National Association of Biology Teachers (2003) has also provided a useful position statement as described in Chapter 10.

Other professional organizations have addressed the use of animals in precollege education. Perhaps the most comprehensive of these is by the American Association of Laboratory Animal Science (2005). It includes statements by science teachers' associations and a copy of the Institute for Laboratory Animal Research (ILAR) guidelines. It also presents specific recommendations on dissection, with less detail than the NSTA statement, and the use of animals in science fair projects.

Finally, for the very common use of pets in classrooms, it is important to note the useful guidelines, i.e., the "IAHAIO Rio Declaration on Pets in Schools" put forward by the International Association of Human-Animal Interaction Organizations (2001). These provide guidance for teachers and other persons and organizations involved in pet programs for schools.

CHAPTER 7

Empowering Teachers to Find Substitutes for Dissection

All the standards, curricula, and legislation aside, teaching of life science and health-related biology takes place in a particular classroom with a group of students led by a teacher. As explained by authorities, "How science is taught ultimately depends on the teachers" (Duschl, Schweingruber, and Shouse, 2007). People who are concerned about the uses of animals in classrooms may wonder why animal dissection persists in seventh grade science, despite largely being phased out in advanced college and professional schools. A more productive approach to this question is to look at these science teachers and their context. Science teachers are known to be overwhelmed by challenges—pupils lack interest in science, teachers may have less than ideal backgrounds in science, schools lack resources, and society lacks interest in education (Matthews, 1998). The societal concerns with science education in the United States are increasing, broad-based, pervasive, and longstanding, and go far beyond the specific use of animals in precollege. But the various factors limiting science learning of students, as discussed in Chapters 4, 5, and 6, also restrain introducing fresh exciting new teaching resources into classrooms, including those that perhaps would supplant partially or completely the interventive uses of animals.

In this chapter we review the rewards and challenges of teaching biology, the use of current resources, and how the precollege educational establishment can update and enhance life science teaching with all these constraints.

REWARDS AND CHALLENGES OF TEACHING

Why do teachers enter the profession? It requires earning an under-graduate degree and then going on with graduate work to earn a teaching credential. At the end, the teachers earn a modest salary. While, like many other champions of the teaching profession we believe that teachers' salaries should be greatly enhanced, something other than money clearly excites students about entering the teaching profession.

Rewards

Teachers, to whom this book is dedicated, enjoy being around students, playing a role in their lives, and participating in their learning process. The contact with students makes teaching rewarding. It takes considerable educational effort and financial investment to become a teacher, involving preparation beyond a bachelor's degree and requiring ongoing in-service workshops and courses. Other professions with a similar educational requirement can promise higher incomes, perhaps fewer daily challenges, and even greater societal acknowledgment. Teachers presumably make their choice of profession based on their enjoyment of working with children and making a positive difference in their lives.

Choosing to become a teacher carries with it idealistic dreams of inspiring children and wanting to see them thrive and develop. Over the course of a year, they come to know some of their students very well. They may maintain contact with some small number of their students, possibly for the rest of their lives, and will have the pleasure of experiencing and being reminded that they made a difference in those people's lives. Even at fiftieth high school reunions, it is not uncommon for former students to re-establish a connection with their favorite teachers of so long ago. Their influence on a student's life choices may still be sharply evident.

Teaching Effectively: Ongoing Preparation

Teachers are invested in their profession and take pride in teaching well. It matters to them that they present lessons in the best way possible. They are alert to new methods, sources of materials, and ideas for freshening up their approaches in teaching. Often they spend some of their own money for teaching materials.

Teachers assume responsibility for organizing their area of subject material, and learning how to teach it effectively. Making learning appealing and fun is a central objective. In the field of science, and especially biology, the subject matter preparation assumes more importance and complexity. Teaching science involves conveying certain principles as well as offering coordinated laboratories that provide practical, hands-on exposure to the same concepts, giving the students opportunities to conduct some of their own experimentation and data analysis. They are always thinking about how better to teach a lesson. Wherever they are, in a grocery store or walking on a mountain path, they are likely to be alert to improved resources to use in the classroom and to try new methods. They sign up for short courses or workshops, hoping to enhance their own teaching.

As a society, we rely on teachers to be inspired and creative in teaching our children. As science teachers, they are responsible for carrying to our children highlights of the exploding knowledge base, enough to motivate our children to want to carry on further study. If they teach biology, they know that for many of their students this is a last formal opportunity to learn about their own bodies and become knowledgeable for managing their own health care and participation in current society.

Increasing Mastery of Teaching and Subject Matter

Teaching precollege in the sciences poses its own moving-target challenges. Even for a university scientist working in a rather finite area, "keeping up" with new developments is an endless task. So, for precollege teachers, who may see over 150 students per day in classes representing several different preparations, it can be very challenging to keep up their mastery of the subject material. Unlike elementary teachers who can concentrate on how best to reach each student largely with material that the teacher long since has overlearned, intermediate and high school science teachers must grapple with complexity of the expanding material. In some cases the teachers never set out to teach biology, but were drafted in from other areas, such as physical education or English, to also teach a few periods in the sciences.

Challenges

Upon entering into the classroom to become a teacher, one can expect some surprises. There always are scheduling adjustments to be made in who is teaching what class due to illnesses, the enrollment of fewer or more students than expected, less classroom space than anticipated, or need to float between different classrooms throughout the day. Administrators are faced with choices to match between the teaching staff that is available, classroom space, time in the day for classes, and number of children who are coming to school to learn. Administrators make the best choices they can with the resources available, sometimes drafting someone to teach a class because they know that teacher has the spark to make the assignment work, even though the teacher's experience does not suggest this capability.

Busy Schedules and Varying Facilities

If the teacher's teaching schedule shows up as six 45-minute periods of teaching 36 children per crowded classroom—over 160 faces per day— plus one preparatory period and lunch, that sets some limits on what can be accomplished in laboratory experiences. Having as many as four separate preparations of material, often the teacher must shift gears during a short passing period to change the laboratory demonstration items that are on display.

Given the requirement to offer laboratories, the type of laboratory classroom and available technology impose clear constraints on what the teacher can do. Some intermediate schools have outstanding laboratory facilities, either within the regular classroom, or as a separate venue. With laboratory tables, students can work in small groups doing observations, gathering data, analyzing data, and writing up results.

Most teachers have access to some computers that can be employed for teaching. Based on our conversations with teachers and focus groups consisting of teachers, the technology can range from a fully equipped computer classroom with each student working on-line and a good technology support staff, to a single computer with projection capabilities available to the teacher for a didactic presentation, to projecting Web site material on an overhead projector, or even just converting Web site material to hard copy handouts.

Many school districts have given a priority to establishing a computer classroom permitting individual access to computers. While one might have anticipated that this could routinely be used to deliver software experiences in biology more effectively to the students, in reality these facilities are commonly so heavily scheduled that it becomes impractical for a teacher to plan for the computer lab to be part of the regular instruction and lesson planning. Teachers instead tend to employ the computer lab as a resource for special projects or individual interest of students.

A paradox is that in some communities almost every student has a computer at home, and yet, this seems almost irrelevant to instruction because teachers cannot require use of these home computers unless every single student has this access. One solution that is common at affluent schools is to guarantee Internet access at school after-hours for the few students who lack access at home.

Expertise in Subject Matter

While there is a need to cover the various courses in a curriculum, the human resources available can pose a challenge. One physical education coach may have minored in biology and be comfortable with a science teaching assignment, but what happens if no one like this is available, and the schedule is short of a science teacher for two-class periods? Real-world situations like this result in someone being drafted to teach a course, feeling ill-prepared, and really having to scramble to do the teaching. Teachers assigned to teach a topic that is far afield from their chosen area of specialization are likely to feel stressed and ill-prepared. It is easy to imagine a teacher's discomfort when accepting a new course assignment and only having the time to stay a few pages ahead in the textbook. Teachers have little recourse for avoiding this discomfort if asked to pitch in on a last-minute assignment when the regular teacher for some reason is unavailable.

Even when the teaching assignment for a particular year is entirely within the preparation area of the teacher, the subject matter in biology constantly expands and requires continued effort to understand. Textbooks, laboratory exercises, and computer software constantly introduce new material and discoveries, some of which are new since the teacher finished school. Teachers are accustomed to attending workshops, conferences, and continuing education courses to keep tuned up on the newest developments, to learn about new technologies, and how to integrate this information and technology into classroom instruction.

Testing to National and State Standards

In recent years, the concepts expected to be learned have been clearly articulated in the standards and often presented with more detail in state frameworks for each subject. Rather than attending to the national standards, teachers in California tend to focus on the more rigorous California state standards and accompanying detailed frameworks for their teaching in biological science and health.

Some clear criteria are suggested in the standards, spelling out important concepts to be presented to students. Teachers have the job of being creative in finding ways to impart and get these lessons across to students. However, the students have a wide range of abilities for understanding the textbooks and other materials. Many are dealing with English as a second language, have a low reading level, or limited linguistic skills in general. The complex concepts of cell and organismic biology may be extremely difficult to work with in classrooms, especially for students with limited English communication skills.

Scheduled tests to assess how students are doing in math and science, along with other subjects, are part of the annual calendar, and the results are published in local communities and analyzed nationally for comparisons internally and with other countries of the world. In the sciences, the national and state standards set the objectives for the material to be tested; students are expected to have mastered the concepts presented in the standards. For the teacher, the examinations tell the tale of whether their teaching efforts have resulted in the expected tested outcomes. Instruction must be oriented toward assuring that the students are familiar with the subject matter described within the standards. The results from a variety of required tests for students make front-page news, showing how we compare with other school districts, other states, and other nations, as well as with performance in previous years. Results of students individually taking SAT tests also are analyzed to monitor students' level of preparation.

The testing intrudes on class schedules, sets an expectation of the material to be taught, and pressures teachers to focus on the test performance of the students. One estimate is that even without specifically preparing students for the testing, about seven instructional days are used for the California STAR testing of fifth graders, 4 percent of the school year, and slightly less for sixth-grade tests (Bill Storm, science teacher, private communication, 2007).

No Budget for Resources

Decades ago in the 1960s, laboratory supplies for science courses were generally part of a regular budget. Some laboratory equipment was purchased, some was on hand at the school resource center, while other resources could be ordered as needed. This somewhat stable arrangement has shifted with the closing of the resource centers for science laboratory materials and curricular libraries with instructional texts, and the generally shrinking budgets for operating laboratories. The challenge to acquire essential equipment and supplies for laboratories can help explain the lack of effective laboratories that may not measure up to the goals of the educational standards. The lack of budgetary support makes it more difficult for teachers to fulfill their mission of inspiring students, and also makes it more likely that teachers will be criticized for poor performance of students, which, of course, is demoralizing for everyone involved.

WHERE DOES DISSECTION FIT IN?

The various challenges outlined above contribute to creating barriers that diminish the likelihood that teachers will mainstream promising alternative resources to dissection in their classrooms rather than using dissection. We have previously emphasized the problematic issues of the topic of dissection not being directly addressed in curricular materials and frameworks, the diminished role of financial, technical, and resource center support in school districts for equipping and supplying laboratories, and the difficulty of locating informative and motivating materials for teaching biology (Hart, Wood, and Weng, 2006). As we see the presentations of almost all areas of biology being regularly updated in cell biology, genetics, body homeostasis, and ecology, the material taught by dissection remains somewhat stuck in the past.

The time has come to address this issue and present some perspectives on how dissection-related subject matter can be made available to teachers in an accessible and affordable fashion. The topics covered by dissection include gross anatomy—where parts of the body are located, the visual images of organs and organ systems, systemic blood circulation, the entire intestinal system, and relation of organs to other organs.

Prior to a discussion of ways in which the subject matter of dissection may be taught without dissection, it is useful to discuss the use of animals in the classroom. There clearly is a place in the classroom for animals. Many teachers enjoy animals themselves and are drawn to classroom pets as a method of facilitating a warm environment where children can learn some nurturing skills while also learning about biology. Most of the experiences with animals offered to children in these years involve animals as whole organisms, although teachers also make use of preserved specimens. In one California study, elementary teachers commonly incorporated animals into teaching via films, story writing, book reports, puppets, non-living specimens, and classroom guests from zoos or museums, as well as live animals kept as "classroom pets" (Zasloff, Hart, and DeArmond, 1999). Teachers believe that having live animals in the classroom helps teach responsibility and respect for living things. The animals stimulate interest and motivation, and the students learn from the experience about the behavior and habitat of animals. Teachers are generally very conscientious with the challenges of providing animal care over weekends and vacations, and problems with noise, cleaning, liability, or welfare. Not surprisingly, the animals kept in classrooms tend to be small, including fish, guinea pigs, rats, mice, birds, and snakes. Many animals are brought in as visitors to the classroom by students, teachers, or visitors from a museum. Dogs and cats are common as visitors. Teachers are also resourceful in using non-living specimens that they kept or borrowed for the classroom, including bones, pelts, feathers, nests, fur, owl pellets, and cow teeth, as well as more conventional educational models or formalin specimens.

Animals Personified in Literature and as Family or Classroom Pets

The children's literature is dominated by stories and books about both fanciful and realistic animals. Figures such as Donald Duck, Mickey Mouse, and Babar, the elephant, are characters resembling children in animal bodies, which makes them somehow more endearing, engaging, and approachable. More realistic animals such as Lassie or Black Beauty offer vicarious experiences involving real animals that could actually happen. Currently about 65 percent of American families have a dog and/or cat. Thus, while children are exposed to cartoonish and realistic animals in literature, at home they typically are encountering actual animals and establishing relationships with them (Figure 7.1).

The pets many teachers keep in their primary-grade classrooms offer opportunities for children to care for, feed, and interact with the pets (Figure 7.2). They may provide exposure to some animals during field trips to nearby science museums (Figure 7.3). While they may welcome and value the pets in their classrooms, most of these teachers majored in English or

Figure 7.1. Even young babies approach dogs and prefer them to mechanical toys. By the time they enter school, most children have had some experiences with dogs and cats that have shaped their attitudes and feelings about them, which in turn influence their views of dissection. (Photo: Joan Borinstein)

other liberal arts, and they often do not feel well prepared in science. Subject matter pertaining to science and health may be given little emphasis, depending on the interests and preparation of the particular teacher. This pattern that is fairly common in the primary grades can also persist into the intermediate grades, at least with some teachers. As mentioned, the conscientious use of classroom animals that are residents in the classrooms, or brought in to visit, is a laudable and valuable aspect of teaching biology and not likely to be replaced by some "higher tech" substitute.

Biology in Intermediate and Secondary Schools

Beginning in intermediate school, the emphasis shifts from the informal use of live animals to learning about the internal structure and physiology of humans and animals, and children are given fewer experiences with the whole animal. The educational science standards emphasize attention to cell biology, genetics, homeostasis of body systems and organs, and ecology.

As children begin their intermediate grades, they typically change classes for science and have a science teacher who has some special preparation in this area. Although the technology revolution may set a high expectation for engaging materials, it can be exciting for students to begin participating in laboratories.

Figure 7.2. Attention-grabbing pets are common in classrooms throughout the elementary grades. Many teachers line their rooms with plants and small animals, such as rats, mice, or guinea pigs. Children inevitably have pleasurable interactions with these pets prior to being confronted with dissection in intermediate and high school laboratories, sometimes involving the same species. [Harcourt Index]

As we discussed in Chapters 2 and 3, intermediate and high school biology courses adopted dissection to provide students with hands-on experience with organology and anatomy and to engage them in the subject matter. Despite major advances in teaching topics such as genomics, cell biology, and computer literacy, laboratory use of dissection remains almost unchanged in intermediate and secondary education. This status quo is reinforced by commercial firms that provide catalogs for conveniently ordering laboratory materials, including dissection specimens. Despite criticism about the numbers of animals being dissected (Balcombe, 1996), advocacy has continued from some teachers for dissection of cats, fetal pigs, or pithed frogs (e.g., Kline, 1995). In this book we offer some guidelines for alternatives for teaching subject matter covered by dissections that would increase the efficiency of learning, add some excitement, and motivate students to learn more.

The teachers of high school biology that retain the traditional pattern of offering dissection of an animal may seek new and improved resources, considering especially those that are easily accessible. But with almost no budget for their laboratories, they are left to improvise and scavenge when designing the laboratories. The disappearance of resource centers that supplement biology materials within school districts isolates teachers as they

Figure 7.3. A python over 4 feet long greets visitors at Explorit Science Center in Davis. Local science centers, such as Explorit, offer convenient sites for field trips that are nearby and on a manageable scale for an interactive tour lasting about an hour. Contact with atypical animals is a favorite activity for many children. [Courtesy of Explorit]

approach this quandary. The overriding motivation of teachers is to stimulate their students to enjoy learning. Dissection remains a favored avenue for providing at least one engaging and memorable experience to students.

Criticism of students' performance in science often focuses on teachers, rather than considering the constraints teachers face. Balcombe (1997) has presented some barriers against acceptance of alternatives in teaching, suggesting that: (1) some teachers are resistant to change, (2) shifting to alternatives requires investing time and money, (3) information on alternatives is not easily accessible, and (4) the quality of alternatives varies. Recently, Balcombe (2001) has made a direct case for adoption of alternatives to dissection. Taking the teachers' perspective, in this chapter we highlight challenges impeding rapid adoption of alternatives to dissection. Within this context, we will recommend the production of Web-based teaching resources to address these barriers and improve instruction in biology laboratories, as further described in Chapter 10.

Attitudes and Practices of Teachers

Despite its rare mention in the educational literature, dissection is commonplace in middle- and high-school level biology. In one U.S. study of almost 500 responding teachers, 79 percent of teachers used dissection and 72 percent believed it was an important part of the curriculum (King, Ross, Stephens, and Rowan, 2004). A majority of teachers, 55 percent, disagreed that available alternatives were as good as dissection for teaching anatomy and physiology. Their primary reason for using dissection was the appeal of offering hands-on work to students. Among those using dissections, only 18 percent used cats.

With relatively little information available from the United States, a few studies from Australia and the United Kingdom are also informative. In a survey in Australia, information was received from 34 schools (Smith, 1994). At these schools 16 species of animals were used for dissection, with use of rats reported by all respondents. Cost was reported as the major factor limiting dissections in these schools, with animal welfare and the teacher preferences as secondary considerations. The teachers believed that dissection should be restricted to children ages 16–17, which was the age of most students doing the dissections. Activities or observations were commonly conducted with live animals, largely invertebrates, as reported by 85 percent of the respondents. All but four of these 34 teachers supported animal dissection and all but three favored studies of living animals in the classroom. Despite the strong support for dissection among this group of teachers, 19 schools reported having substantial problems with dissection; only seven reported having no problems. Some students reported feeling nauseous or squeamish. Concerns with lack of respect for the animals, and a health concern with formalin were also mentioned.

A similar study in England surveyed 28 schools in an urban area with students in the 11–18 years range (Adkins and Lock, 1994). A majority of teachers said that while recommendations of several governmental agencies had no effect on their use of animals, no agency encouraged the use of animals. The various agencies were Her Majesty's Inspectors of Schools, Department of Education and Science circulars, the Assessment of Performance Unit national testing, the National Curriculum, and recommendations by Universities Federation for Animal Welfare. More than one-third of the teachers held opinions that discouraged the use of animals in their teaching. Public and student opinions were viewed as potent factors discouraging the teachers from using animals. Around half of the teachers believed that the most potent factors discouraging the use of animals were the facilities, costs of animals and food, the availability and training of technicians, and the lack of teachers' time.

In our focus group discussions with teachers, we have learned that teachers are unprepared to know what will be gained by offering dissection and other laboratories to students. Perceived as solely for affluent schools and being unrelated to standards or clear outcomes, the teachers' commitment to an outcome is undefined. In its uneven use, some teachers offer inappropriate use of dissection, such as a forty-five-minute laboratory for dissecting fetal pigs, and others find it distasteful and avoid it (Bill Storm, science teacher, private communication, 2007).

Past Experiences of Elementary and Secondary Teachers

It seems likely that choices of teachers to use certain methods or set up particular laboratories for a lesson would be influenced by the teacher's related experiences and knowledge. For teachers, their family attitudes and culture and specific lifetime experiences with animals may influence the extent to which they become engaged in increasing animal-related activities within their classrooms. One would expect a teacher's prior experience with dissection to affect their choices as to having students perform dissection in class and how they respond to students preferring to decline participation in dissection. Working with animals or animal specimens requires an emotional comfort level as well as a feeling of proficiency with the relevant subject matter. While no data are related to this concept, it is known that the likelihood of pet ownership by adults is related to whether they had pets as children (Kidd and Kidd, 1997).

Convenient Resources

Teachers assemble their own lessons and laboratories, and draw from what is available to them that also fits within their scheduled time and seems appropriate for the students in the classes. With science teachers being fairly beleaguered in their schedules, classroom and laboratory preparations, and paper grading, having resources near at hand without great effort seems essential. Convenient resources are important, since all areas of science teaching require laboratory materials. Supporting them in having access to a wider array of appealing and effective resources and teaching opportunities assists them in engaging and motivating their students. To this end, Table 7.1 provides a list of organizations offering information on alternative teaching resources that may be useful for teachers wishing to explore available resources. If teachers must go through extensive hoops to acquire resources, or must plan a long while ahead to obtain them, the probability of them using those particular items is much less. As presented in Chapter 11, a vast number of teaching resources are extensively cataloged in the NORINA database, but it could be a formidable obstacle to have to select a new resource for a lesson without actually working with it, and then to have to identify available funds to use and order it by whatever process is acceptable within a school district.

Some items are also available for loan from several loan programs that are listed in Table 7.2. This smaller pool of resources is selected as being appropriate for K-12 instruction, and in a sense the shorter list may be more feasible for teachers to deal with. Increasingly, the loan programs focus on supplying resources that are linking to science education standards, and on providing fast and convenient service, even using expedited delivery services to meet classroom needs, with only a fee for the return shipment. Using resources provided by the loan programs, although still requiring

Table 7.1 Organizations Providing Information on Alternative Teaching Resources

Title and URL	Producer	Scope
American Association for Laboratory Animal Science (AALAS) http://www.aalas.org/	American Association for Laboratory Animal Science	AALAS is a forum for the exchange of information and expertise in the care and use of laboratory animals. Interested in humane care and treatment of laboratory animals and quality research that leads to scientific gains that benefit people and animals, they support programs, products, and services to that end. They have developed resources relevant to teachers and students using animals in the classroom.
Caring for Animals: The Use of Animals in Precollege Education http://www.aalas.org/pdf/ precollege_education.pdf		
Position Statement on Humane Care and Use of Laboratory Animals http://www.aalas.org/ association/position _statements.asp#precollege		
Kids 4 Research - Caring for Animals in the Classroom http://www.kids4research. org/animals.html		
Association of Veterinarians for Animal Rights (AVAR) Alternatives in Education Database: http://www.avar.org/ alted/	Association of Veterinarians for Animal Rights	The Alternatives in Education Database contains thousands of entries of alternatives for many levels of education.

(continued)

Table 7.1 (*continued*)

Title and URL	Producer	Scope
		To locate a particular alternative, type in a key word or phrase in the blank box. The advanced search option allows searching for an alternative by species, medium, discipline, educational level, or any combination of these categories.
Animalearn http://www.animalearn.org Animalearn: Animals, Ethics, & Education—Science Bank http://www.animalearn.org/sciencebank.php	American Anti-Vivisection Society (AAVS)	Animalearn is the educational division of AAVS, an animal advocacy and educational organization. Focus is to foster an awareness and a respect for animals used in education. Their goal remains to eliminate the harmful use of animals in education and to work to assist educators and students find effective nonanimal methods to teach and study science. Animalearn created the Science Bank, a collection of life science software and educational products available for educators and students to borrow. Animalearn also provides humane education curricula and materials free of charge for educators and students.
Animals in Education http://www.navs.org/site/PageServer?pagename=ain_edu_education_main	National Anti-Vivisection Society (NAVS)	NAVS supports the Animals in Education section, which includes resources and materials on alternatives to dissection in the classroom, the BioLeap Lending library of life sciences materials, science fairs, and a list of relevant links.

(*continued*)

Table 7.1 (*continued*)

Title and URL	Producer	Scope
Animals in Education http://www.aavs.org/ education01.html	American Anti-Vivisection Society (AAVS)	The Animals in Education section of the AAVS provides information on Animalearn, dissection, humane education, student choice, and resources.
Computer-Aided Learning in Veterinary Education (CLIVE) http://www.clive.ed. ac.uk/	Learning Technology Section, College of Medicine & Veterinary Medicine, The University of Edinburgh	CLIVE: Computer-aided Learning In Veterinary Education is a consortium of six U.K. veterinary schools. They are working to make Computer-Assisted Learning (CAL) an established and expanding feature of veterinary undergraduate education in all subjects of the veterinary course.
Dissection Alternatives, Physicians Committee for Responsible Medicine (PCRM) http:// dissectionalternatives. org/	Physicians Committee for Responsible Medicine	PCRM supports alternatives to teach the concepts of anatomy and biology, using computerized techniques to allow students to explore human or animal anatomy. The Web site provides information and access to specific alternatives as well as material on student choice.
European Resource Centre for Alternatives in Higher Education (EURCA) Alternatives Database http://www.eurca.org/ http://www.eurca.org/ resources.asp	European Resource Centre for Alternatives in Higher Education, Utrecht University and University of Edinburgh	EURCA promotes the use of alternatives to using animals in higher education, with the goal to provide a mechanism for effective dissemination of information about alternatives to using animals in education. The Alternatives database provides full descriptions of available priced alternatives as well as reviews commissioned by EURCA, user comments, and other reviews and descriptions.

(*continued*)

Table 7.1 (*continued*)

Title and URL	Producer	Scope
From Guinea Pig to Computer Mouse: Alternative Methods for a Humane Education http://www.interniche.org/book.html InterNICHE http://www.interniche.org/	International Network for Humane Education InterNICHE	InterNICHE is a network of students, teachers, and animal campaigners, focusing on animal use and alternatives within biological science, medical, and veterinary medical education. They offer an Alternatives Loan System, a library of products available for free loan anywhere in the world as well as literature, support, and advice for teachers and students. The book, *From Guinea Pig to Computer Mouse: Alternative Methods for a Progressive, Humane Education* suggests alternative tools and approaches within biological science, medical and veterinary medical education. It is freely available on the Web and is published in several languages.
Historical and scientific perspectives on why animals are used in research (OSERA) http://www.osera.org/buspcatMXBZ74GU.html	Ohio Scientific Education & Research Association (OSERA)	OSERA's Web site discusses the book *Cattle, Priests and Progress in Medicine* by Calvin W. Schwabe. It provides a detailed description of the evolution of medicine and especially veterinary medicine.
ITG Whole Frog project http://www.lbl.gov/ http://froggy.lbl.gov/virtual/	Lawrence Berkeley National Laboratory	Berkeley Laboratory conducts research across a wide range of scientific disciplines and in the use of integrated computing as a tool for discovery. Believing that computers might contribute to the field of

(*continued*)

Table 7.1 (*continued*)

Title and URL	Producer	Scope
		teaching and anatomy, they developed the award-winning virtual frog.
New England Anti-Vivisection Society (NEAVS) Ethical Science Education Coalition (ESEC) http://www.neavs.org/esec/index.htm Loan Libraries, Databases, Publications http://www.neavs.org/esec/alternatives/alt_dissectionalternatives_loan_libraries_101501.htm Alternatives to Dissection; ESEC Resources http://www.neavs.org/esec/alternatives/alt_index.htm	New England Anti-Vivisection Society	NEAVS advocates for animals through public outreach efforts, publications, education programs, and legislative initiatives and litigation. The ESEC is the educational affiliate of NEAVS, offering instruction and resources for education that do not require students to dissect an actual specimen. ESEC works with educators, legislators, parents, and students to promote humane science education at all levels of education and professional training. They offer assistance in the selection of alternative teaching tools and the development of lesson plans for teachers, information on funding available for alternatives, including loan libraries and pilot program grants, and Professional Development Points (PDP), and training for teachers on the use of alternatives.
The National Association for Humane and Environmental Education (NAHEE) http://www.nahee.org/	Humane Society of the United States (HSUS)	NAHEE is the youth education division of the (HSUS), working in education to educate young people about animals and to provide teaching materials, professional development, and other support to teachers and humane educators.

(*continued*)

Table 7.1 (*continued*)

Title and URL	Producer	Scope
NORINA http://oslovet.veths.no/ fag.aspx?fag=57 http://oslovet.veths.no/	Norwegian Inventory of Audiovisuals (NORINA), Norwegian Reference Centre for Laboratory Animal Science & Alternatives	The NORINA database contains information on over 3,800 audiovisual aids that may be used as alternatives or supplements to the use of animals in teaching. The information in the database has been collected from 1991 until the present.

that teachers plan in advance, ducks the problem of cost and increases the array of resources.

Having a comprehensive Web-based resource that could be drawn from for teaching certain subtopics of particular interest would offer resources at one's fingertips to be accessed exactly when needed, and with minimal special effort. Many highly motivated teachers already make use of the patchwork of isolated effective teaching tools that currently are available on the Web, and even share their favorite resources with other teachers (e.g., Storm, 2007).

POSSIBLE STEPS

Ultimately, it is teachers who deal with the dissection dilemma on a daily basis. One major sign of progress would be to have a more inclusive treatment of dissection and laboratory methods within the educational literature on teaching biology. A more open discussion of the various issues, challenges, and dilemmas would lead to improved solutions for everyone.

Targeted Professional Development

Virtually every analysis of the crisis in science education addresses the professional development of teachers as a central issue that needs to be addressed. For example, in one study even with major changes in the science curriculum and science education, no changes appeared in students' perceptions of and attitudes toward science, leading to a conjecture that more emphasis is needed on the teachers' role and teaching style if an educational change is to be achieved (Ebenezer and Zoller, 1993).

Table 7.2 Loan Libraries for Teaching Resources

Title and URL	Producer	Content
Alternative Resource Center http://www.neavs.org/ esec/alternatives/alt_ dissectionalternatives_ loan_libraries_101501. htm#esec Alternatives Loan Library http://www.neavs.org/ esec/alternatives/alt_ Loan_Library-Page-1. htm	New England Anti-vivisection Society, Ethical Science & Education Coalition (ESEC)	Dissection Alternatives resources include lists and access to loan libraries, databases, and publications. ESEC's Resource Room contains information on more than 400 books, 200 videos, and dozens of models and computer programs available for use.
Alternatives Loan System (InterNICHE) http://www.interniche. org/alt_loan.html	International Network for Humane Education	The InterNICHE Alternatives Loan System is a library of multimedia CD-ROMs, videos, models, and mannequins, covering fields such as anatomy, physiology, and surgery. Teachers and students from anywhere in the world can borrow items from the Alternatives Loan System.
Animal Alternatives in Education http://www.navs.org/ site/PageServer? pagename=ain_edu_ education_main BioLEAP Lending Library of Life Science Materials http://www.navs.org/ site/PageServer? pagename=ain_edu_ dissection_loan_program	National Anti-Vivisection Society (NAVS)	BioLEAP Lending Library of Life Science materials is included among the resources gathered in the Animal Alternatives in education section at NAVS. The extensive library of dissection alternatives is available for free loan to students, teachers, and school boards. The Library of Alternatives includes dozens of high-relief, three-dimensional plastic models, computer software programs, color transparencies, and videotapes for students from elementary through the postgraduate level.

(*continued*)

Table 7.2 (*continued*)

Title and URL	Producer	Content
Animalearn http://www.animalearn. org/home.php The Science Bank http://animalearn.org/ sciencebank.php	Animalearn, American Anti-Vivisection Society (AAVS)	Animalearn provides a host of resources for educators and students for educational levels K-12, college/university, and veterinary/medical. Resources include humane education curricula and educational kits covering issues ranging from dissection to product testing as well as a variety of books, brochures, and videos. The Science Bank is Animalearn's lending library, and offers over 400 products, each available in multiple quantities to outfit entire classrooms. The humane education curriculum, Next of Kin, is aligned to both national and Pennsylvania state standards to meet teachers' needs for math, science, and ecology.
Humane Education Loan Program (HELP) http://www.hsus.org/ animals_in_research/ animals_in_education/ humane_education_loan_ program_help/index.html	Humane Society of the United States (HSUS)	The Humane Education Loan Program of the HSUS offers middle schools, high schools, and universities alternatives to be used in the classroom. All of the loan materials listed are available for the cost of return postage.

Easier Access to Effective Resources

The curricular requirements for developing some laboratories in high school biology that conform to standards are relatively simple and straight-forward. In crafting such a solution, it is critical to recognize that teachers have limited time available for laboratories on these topics; the solution should not require extensive resources to meet their needs. This is where making software available for selected topics could improve basic education in biology. Even five outstanding laboratories produced in software and made freely available on the Web, could revolutionize biology laboratories in many classrooms, provided they had the required properties of flexibility in use, content parameters, and reading level.

Why is improved and exciting teaching technology not currently being delivered to intermediate and high school biology? Formerly, school districts provided more active mentor support and assistance to teachers and maintained science curriculum centers that loaned laboratory materials and curricular guides. Today, computers are widely available, while resource centers and curricular libraries have been closed. Economic challenges limit the resources that most school districts can provide. Many teachers feel stretched with content requirements and lack of resources, while seeking to inspire their students. They accumulate materials over the years and use their own money to enhance laboratories. Teachers in focus groups in 1996, 1997, 2004, and 2007 have indicated to us that current resources do not address their needs for intermediate and high school biology and physiology: they are expensive, do not fit the curriculum, and are visually inferior. Teachers lack well-integrated and accessible materials to draw from to meet their needs, technology capabilities, and schedule constraints. However, with the explosion of science museums, the Internet, and the commonplace at-home computers, much more can be done to incorporate informal with formal education. Schools can make better use of the learning opportunities their students experience outside the classrooms and thus complement the laboratories.

CHAPTER 8

Students and the Culture of Dissection

Whenever students are presented subject matter pertaining to animals, for them this tends to call up their personal experiences with animals as well as whatever the content of study may be. Most families have companion animals at home as well as in their neighborhoods, and in more upscale communities, it seems that even more families have animals at home. Many children naturally come to have a personal relationship with several animals as individuals. Thus, the study of animals is, in a way, like studying a family member or a close friend, and not at all like learning about the planet Jupiter, or the molecular composition of cell membranes, or the power of gravity. With science in general, the topic can be a subject of in-depth study and data collection for further learning. However, with animals, a background of emotional experience places the subject matter in a different framework.

Understanding something about the students' attitudes and knowledge concerning animals and life processes is essential when setting curricular goals. As stated by Ausubel (1968), "The most important single factor influencing learning is what the learner already knows." Thus, a central idea is to ascertain what children know, and adjust the teaching accordingly, so that they can enhance and add to what they already know and understand, a central concept in constructivism (Matthews, 1998). Teaching and learning approaches based on this theoretical position consider knowledge as a human construction, in which students' conceptions and beliefs about knowledge, learning, and teaching form the framework for their learning, and are referred to as constructivist. Essentially, students construct their own meanings from words or visual images that they hear or see, and go on to construct explanations of their experiences (Treagust, Duit, and Fraser,

1996). With a view that students actively construct their own knowledge, it becomes essential to ascertain the nature and extent of learners' knowledge and experience in the area of interest.

ATTITUDES AND EMOTIONAL EXPERIENCES REGARDING ANIMALS

Because a student's reaction to dissection is going to be influenced by emotional experiences regarding animals, observations of toddlers reveal their distinctive responsiveness to live pets, more than mechanical ones, from a very young age (Kidd and Kidd, 1987). They prefer interacting with the live dog, and respond to a live pet in a more lively way. Many young children (age ten) regard their pets and neighborhood pets as best friends (Bryant, 1985). How they regard animals as they get older, and their probability of acquiring animals as adults, is correlated with their family pet-keeping history, that of both their parents and grandparents (Kidd and Kidd, 1997).

In their experiences with cats and dogs, children rate them among their "top 10" most special relationships (McNicholas and Collis, 2001). These animals provide significant support functions to the children, ranking higher than certain kinds of human relationships. They provide comfort, support esteem of the children, and serve as confidantes for secrets. These special relationships with cats and dogs confer on them a specialized role. Additionally, a significant correlation has been reported for pet ownership and empathy (Daly and Morton, 2006).

A substantial concern in educating children involves methods for fostering empathy, and the possibility that encouraging children's kindness and caring toward animals may be useful and generalizable to human-directed empathy (Ascione, 1997). A growing body of evidence shows that an enhancement of attitudes toward animals generalizes to human-directed empathy, especially when the quality of the children's relations with their own pets is considered (Ascione and Weber, 1996).

Children's experiences with animals extend beyond just dogs and cats, and pet owning children are more likely than nonowners to have a wide range of experiences with animals, such as visiting zoos and reading books about animals (Kidd and Kidd, 1990). In a study of school children in England, 80 percent of the students had current experience with keeping pets at home (Lock and Millett, 1992). Typical involvement with animals in their spare time differed between girls and boys, with the most common activities being almost a quarter of the girls engaged in horse riding, and more than a third of the boys experienced in fishing.

Although many children find their lives enriched in positive experiences with companion animals, by no means is this the universal outcome. Children in problem-ridden neighborhoods who are at high risk for academic problems often have been threatened by dogs and remain very fearful of

Figure 8.1. Children whose prior exposure has been to unfriendly dogs, rather than friendly ones, have an opportunity to work in pairs and train a shelter dog at centers such as the SPCA-LA Tender Loving Care Program. Close supervision by a trainer assures that the child makes progress and is not endangered. Through their experiences with the dog they are training, the children's fear of all dogs shifts to a more realistic respect, and an affection for certain dogs whom they know. [Courtesy of SPCA-LA]

them. The TLC (Tender Loving Care) program at SPCA-LA (Society for the Prevention of Cruelty to Animals, Los Angeles) pairs up children with shelter dogs as part of a violence prevention program, and assists them in learning to train the dogs, thereby lessening the children's fear of dogs while increasing their knowledge about dogs (Zasloff, Hart, and Weiss, 2004). Examples are shown in Figure 8.1.

Attraction and Engagement

Much of the appeal of having animals in classrooms and providing exposure to animals, and even providing exposure to preserved specimens, is that they engage students in learning partially because of their prior experiences with animals. Studies of student engagement in learning tasks report that children who are engaged show positive emotions, including enthusiasm, optimism, curiosity, and interest, in contrast with disaffected children who are passive and withdrawn, give up easily, and may be bored (Skinner and Belmont, 1993). Formal methods are available for assessing student engagement rates during various tasks (Chapman, 2003). Unfortunately, such technologies have not been applied to studies of classroom laboratories, including those with dissection.

In a conversational analysis of children visiting three types of animal exhibits, a museum, a zoo, and animated dinosaurs and preserved animals, the children's age did not seem to affect the content of conversations, but the older children expressed more emotions (Tunnicliffe, 1996d). Movement was a powerful feature affecting the attention of observers (Tunnicliffe, 1996a). School children visiting a farm had more emotive and affective comments than those visiting the zoo (Tunnicliffe, 1998b). The children especially enjoy the close or even tactile contact with animals. While there were many similarities in the conversations of primary school parties visiting animal specimens in a museum and zoo, the museum visit appeared to present a more structured educational experience for the children (Tunnicliffe, 1996b). In learning zoology, the active behavior seen in a zoo may actually detract from the learning of structural zoology.

During the visits to a natural history museum, the conversational content did not differ when an adult was present, suggesting that either the benefit of having a teacher present is minimal, or else the students were very well prepared (Tunnicliffe, 1997a). This series of studies of children's visits to zoos and museums finding no marked distinctions in conversational content in family visits to the zoo versus school visits to the zoo implies that the educational opportunity is not effectively used during school visits (Tunnicliffe, Lucas, and Osborne, 1997). However, these data are difficult to interpret given that field trips to zoos involve large groups of children, whereas parents taking their children are supervising a much smaller number of children per adult. Conversations during school visits involved some increase in references to animals' body parts and comparisons between animals. There was also a higher level of affective comments during school visits, which may have reflected their communicative patterns with other students.

In a study where ten-year-old children participated in an extracurricular course on the structure and function of animals, they had four instructional strategies: direct observation of animals, watching videotapes, participating in class discussion, and building kits and games (Tamir and Shcurr, 1997). From assessment at the end of the informal course, the attitudes of all students toward animal subject matter were improved. Girls, who initially were less positive to small animals than boys, expressed as favorable an attitude as the boys following the course experience. Girls preferred direct observations and building kits, whereas the boys preferred discussions.

One consistent finding across ages is that girls are more emotional than boys about animals. This widespread result is well known from the work of Kellert (1985, 1996; Kellert and Westervelt, 1983) on attitudes to animals, consistently showing higher ecologic scores for boys and humanistic and negativistic scores for girls. Thus, boys more often than girls wished to learn more about how species coexist and are useful to each other. Girls more often reported that they like pets very much, and that they were fearful of or did not like snakes or spiders. Studies in other countries, as illustrated in Norway (Bjerke, Odegardstuen, and Kaltenborn, 1998), have yielded similar results. Aside from gender, differences in children's knowledge, attitudes, and behaviors toward animals also were noted in Kellert's studies as a function of age, ethnicity, and urban/rural residence.

Tunnicliffe's (1998a) studies of conversational analyses at animal exhibits in England found in comparisons of single sex groups that primary school boys named the specimens more, whereas the girls made more emotive comments. A large probability sample of adolescents in the United States found that the most striking relationship was the difference in attitudes toward animal research on the basis of gender (Pifer, 1994). The author draws some parallels between opposition to animal research and feminist attitudes.

Among undergraduate university students, men were more supportive than women of the use of animals in biomedical research (Hagelin, Hau, and Carlsson, 1999). Female veterinary students at two British universities rated themselves as having higher levels of emotional empathy with animals than did the male students, and the women maintained their high scores on empathy throughout veterinary school, whereas the male students had lower levels in their later years (Paul and Podberscek, 2000).

Providing positive experiences with animals to children, sometimes in a context of humane education curricula, is one avenue for fostering empathy for other living things. In the converse circumstances, where children are exposed to abuse of themselves, other family members, or their animals, they may learn negative behaviors (Ascione and Arkow, 1999). Children who have exposure to younger siblings gain experience with empathy, whereas the only child or the youngest child will seek out companion animals or sometimes elderly people in the neighborhood, and gain experience in expressing empathy in those relationships (Melson, 1988).

Disgust or Apprehension of Students about Animal Use

While we typically hear about children's attraction to animals, we could also focus on the impact of feeling disgust, apprehension, or aversion about exposure to animals in a laboratory context. With dissection, part of the experience is the smell of formalin and the squishy-looking organs in the abdomen. Students may feel a mixed approach-avoidance polarity in facing this experience; it is unpleasant and very intriguing at the same time. The presence of disgust plays a role in making the experience much more memorable, perhaps accounting for so many people being able to recall certain details from their dissecting experiences in high school, whereas most of their other classwork is long forgotten. Students' mixed feelings are illustrated in the quote from a teacher: "In my work with fifth graders dissecting chicken wings, deep empathy is elicited from many students *for the chicken*, and there is often a good deal of resistance from some students because they value the chicken and not because they are squeamish. This increases somewhat when I draw their attention to the sacrificial nature of our use of the chicken wings for anatomical study (from a grocery store), but while their initial resistance may increase, actual participation is increased and squeamishness diminishes. Parents have told me I have created a number of vegetarians with that one lab" (science teacher, Bill Storm, private communication, 2007).

Noted physician–author Oliver Sachs (2001) in his memoir, *Uncle Tungsten: Memories of a Chemical Boyhood,* recalls being pushed by his obstetrician mother to dissect human malformed fetuses. "She never perceived, I think, how distressed I became . . . Though I had taken to dissecting naturally, by myself, with earthworms and frogs and with my octopus, the dissection of these human fetuses filled me with revulsion . . . I saw in these fetuses who . . . I, too, could have been, and this made it more difficult to distance myself, and heightened my horror . . . these precocious experiences turned me against medicine, made me want to escape and turn to plants, which had no feelings." His mother persisted in introducing him to human anatomy and introduced him to his own cadaver at the age of fourteen, a fourteen-year-old girl. Despite these emotionally searing experiences, Sachs still ultimately entered human medicine, so we are left not knowing whether the net effect of his early exposures to cadavers was positive or negative with regard to professional choice.

An ethnographic field study of sixth-grade science classes performing dissection of fetal pigs noted that most students were initially ambivalent and squeamish (Solot and Arluke, 1997). Using various strategies, the students transformed the situation to a positive experience, by accenting the positive value of the experience, joking, and avoiding portions of the dissection. Transforming the fetal pig into a specimen was part of the strategy for managing emotions.

Attitudes toward use of parts of animals or whole animals for dissection in schools were explored in a survey of 468 students in England, among whom 80 felt it was unnecessary, 47 felt squeamish, and others found it cruel (44), wrong (37), or disliked it (10) (Lock and Millet, 1992). This study led the authors to call for more discussion of the students' views about dissection to assimilate a balanced range of different views on the topic. Another study of British secondary schoolchildren found that 73 percent of younger students felt "bad" about carrying out their first dissection, and even at the age of 15 a majority of girls still felt this way (Paterson, 1986, cited in Langley, 1991).

Here is a practical tip for teachers using dissection or other animal-based exercises. Acknowledging and discussing the emotional reactions is useful and may modify some of the more extreme reactions. When dissection tends to evoke moral concerns in some students, offering memorial rituals or funerals to acknowledge the contribution of the animal to education is one method of addressing uneasiness for use of animals. Along these lines, some schools host a funeral to bury the remains of cadavers. Acknowledging the moral stress and providing avenues for learning to deal with it serve students well in learning to deal with other sources of stress in the future.

Recently, the plastinated, sliced, and staged human bodies placed in artistic and athletic poses, in the BODY WORLDS exhibit (2007a), have drawn crowds in cities throughout the world, permitting children and adults to explore all parts of human bodies at a distance of just a few inches. The ancient fascination with the human body, and perhaps especially with reproduction, is evident from early drawings by Mondino (Figure 8.2) and DaVinci (Figure 8.3), among many others. Suddenly, it is possible to see the insides of real human bodies, including a fetus, or a pregnant body in Professor von Hagens' BODY WORLDS. In his words, "The democratization of anatomy that the BODY WORLDS exhibition affords us has primarily been demonstrated by the fact that laypersons have reacted in a completely different way to the exhibition than was predicted by experts. The high number of visitors reflects the general population's *need to know more about the structure and functions of their bodies*" (italics added) (BODY WORLDS, 2002). In fact, the current widespread availability to the public of exhibits with such art work and real human bodies seems to reveal that school policies requiring parental consent to discuss any aspect of reproduction are now an anachronism.

The appeal of intriguing and even disgusting elements of dissection is sometimes exploited. A best selling children's book, *Grossology—The Science of Really Gross Things*, inspired a popular science exhibit, "Grossology: The (Impolite) Science of the Human Body," which promised to delight many kids and grownups (Gathright, 2003), while perhaps arousing moral conflict for others.

Teachers talked about the phenomenon of disgust and such aversions during pilot-tests of the software, *The Virtual Heart*, in some high schools in Sacramento Valley. The image of a mass of worms inside a dog heart was one of the big attractions of the software—eeeewww!, the "grossed-out" students would say, while still being intrigued with this picture. Being *grossed out* is not necessarily entirely a bad thing in educational materials (Nabi, 2002), and effects may differ between genders as described above.

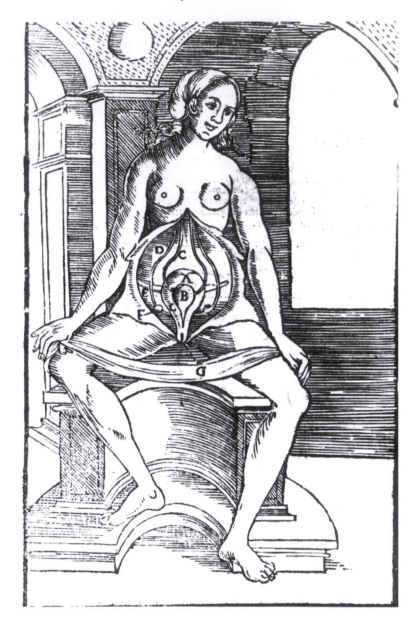

Figure 8.2. A strong interest in women's organs, even in the time of Mondino dei Luzzi (1270–1326), is revealed in this drawing. Mondino's work was based on his dissections of humans and also pigs (Photo: National Library of Medicine)

There is some evidence that disgust is especially salient for women, raising a possibility that disgust could be discouraging some girls from entering the sciences (Fessler, Pillsworth, and Flamson, 2004). On the other hand, one could argue that whether dealing with human or animal biology, part of

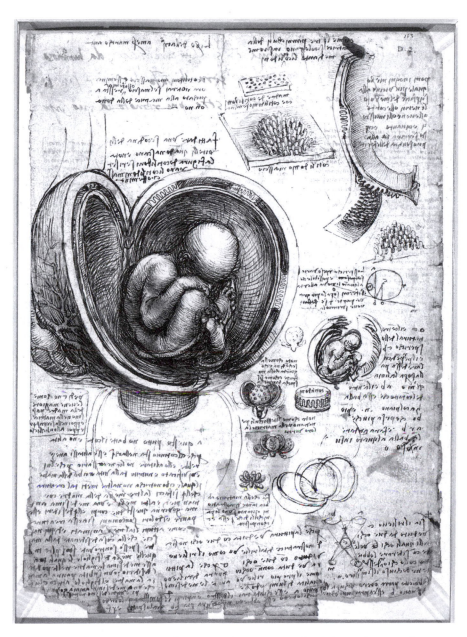

Figure 8.3. This classic image by Leonardo DaVinci, The Babe in the Womb (1511), reflected his knowledge gained from dissection of human bodies, as well as his experience with animal dissection. The image presents the placenta to be like that of the pregnant cow he had dissected, rather than showing the accurate structure of the human placenta. [Courtesy of The Royal Collection © 2006 Her Majesty Queen Elizabeth II]

the educational mission is to move students beyond the grossed-out stage to learning something about biology and anatomy. After all, to be able to use all aspects of biology in managing one's health, and that of family members, requires dealing with body systems without being grossed out.

From the historical review in Chapter 2, one can expect strong emotional experiences to arise concerning dissection of animals or humans in laboratories, even among adults. Medical students in one study articulated the subtleties of the dilemmas they faced, and described their uneasiness regarding the moral implications of using dogs (Arluke, 1996, 2004). However, these feelings were resolved during the laboratory, so that later the students described the lab in very positive terms. Arluke recommends classroom discussions of the moral concerns that students may have, prior to the dissections.

Knowledge of Living Things

Children's awareness of the concept of "living" as it develops with age was studied by Piaget, who found that young children attributed many inanimate objects with life, and only accumulated an adult concept of living around the age of eleven (Driver, Squires, Rushworth, and Wood-Robinson, 1994), the age when some students are making jokes about fetal pigs, even as they are acquiring moral sensibilities. Very young children operate with a naïve theory of biology, in which they may lack a mind–body distinction that is characteristic of our culture even though they have a form of intuitive biology (Carey, 1985). It is argued that these concepts of intuitive biology are changed and reorganized during development (Hatano and Inagaki, 1997). In the naïve biology of young children, human bodily processes as well as nonhuman animals and plants, are topics of apparent concern and importance (Inagaki and Hatano, 2002). Even young children are fairly knowledgeable about these targets of concern and willing to talk about them, including such topics as hunger, illnesses, and reasons for a balanced diet, and they are interested in "reading" picture books about nonhuman animals, visiting zoos, and watching TV programs about nature.

Various studies indicate that by the age of ten, children seem to understand that the body includes several organs that work together to maintain life (Carey, 1985). Children's feelings that only animals but not plants are alive, or that both animals and plants are alive, positions them to begin gaining biological knowledge and understanding biological processes. A study reported that although Scottish children aged 10–15 could provide examples of living things, only 9 percent could correctly classify a list of items as to living or nonliving, and a minority understood that respiration is an attribute of living things (Arnold and Simpson, cited in Driver et al., 1994). Several studies reveal that children slowly acquire an understanding of the more subtle attributes of living things, and often have difficulty recognizing that embryos or certain plants are alive.

How children feel about science in general inevitably plays a role in how they respond to classroom experiences. The period of eight to thirteen years has been emphasized as the critical stage in which children's attitudes toward science can be influenced (Ormerod and Duckworth, 1975). Unfortunately, though, a review presents evidence that children's attitudes to science tend to decline during the intermediate and high school years (Koballa, 1995).

Informal education plays a role in children gaining familiarity with animals. During visits to zoos with family or classmates, children focus on naming the animals and perhaps also note anatomical features such as body covering and number of legs (Tunnicliffe, 1997b). Not surprisingly, children tend to focus specifically on large four-legged terrestrial mammals, and are less likely to classify fish, invertebrates, and whales as animals (Bell, 1981). Even university students are most likely to mention large terrestrial mammals when asked to name five animals, and not to list a more representative assortment of animal life (Trowbridge and Mintzes, 1985). Having a rudimentary understanding of the attributes of living things before the age of twelve positions children to be prepared to develop their knowledge further when studying biological science in grade seven, when it is typically offered.

Concerning students' attitudes about the cognitive or psychological capabilities of animals and their feelings regarding using animals in laboratories, most studies have focused on college-age students. To assess what students think about animals' capabilities, in terms of animals being able to think about what another animal is thinking, a group of over 200 psychology students in the United Kingdom who were applying to college ranked the species, in declining order of cognitive prowess, as primates, dogs, cats, birds, rodents, and insects (Furnham and Heyes, 1993). Females rated the emotional capabilities of animals significantly more highly than did males. In another study of young people (average age, 20 years), higher education and being female were associated with being against use of animals in laboratories (Furnham and Pinder, 1990). A study of psychologists in the United States revealed that most respondents favored using animals in undergraduate psychology courses, with 58 percent supporting this use of animals, yet, 54 percent felt that laboratory work with animals should not be required (Plous, 1996).

Awareness of Death, Experience with Pet Loss

Children begin to have a biological understanding of death around age nine or ten, learning that the body ceases to function as an inevitable biological process (Carey, 1985). Many children experience the loss of a beloved animal during their elementary school years. Whether a family pet at home, an animal at the home of a grandparent, or a neighborhood pet, the death of an animal friend for children is a sobering introduction to the temporariness of life, as with the loss of a grandparent or other family member. It brings sadness, a sense of loss, and a grieving for the individual traits of that animal or person. Having an experience of losing a beloved dog or cat, sets up a child for being ill prepared to see another animal sacrificed, presumably for their education, which may lead them to feel responsible for the death of that animal.

One sensitive point not usually discussed is that cats are among the animals frequently dissected in precollege laboratories, and they also are special animal companions for many children. The emotional reactions of some students against dissection are likely to be exacerbated if they have been particularly attached to cats. For such students, the idea of simply assuring the child that the animal has had a painless death skirts the more important question of the value and quality of the animal's life.

Ethical considerations of using animal bodies in laboratories are a serious concern in higher education, but are little explored in precollege. One survey of high school biology and social studies teachers in Australia, Japan, and New Zealand found very divergent views from the three countries on animal experiments, bioethics, and whether any guidelines were in place (Tsuzuki, Asada, Akiyama, Macer, and Macer, 1998).

Based on experience, one science teacher expressed the sentiment: "As medical schools now provide guidance in the way human cadaver materials are handled with a degree of reverence, schools entrusted with the sensibilities of even more impressionable children provide no such training for their teachers as they set about exploiting animals that for many children represent beloved pets and companions.... Many (teachers) are aggressively dismissive of a need for ethical and reverent handling of animal materials, believing that children should be expected to be as casual with the materials as they, and that any concession to these ideas opens to the door to the elimination of dissection as a practice.... I have always found that ethical and sensitive student preparation in approaching any dissection, be it a clam or chicken wing, and an atmosphere of gratitude, has always promoted a healthier and richer learning environment (respect for the materials and focus on the task, for instance), thus a better outcome for students" (Bill Storm, private communication, 2007).

For some students, knowing that animals have given their lives for their education becomes a significant burden. Although uncommon in the United States and Canada, there are examples of formal ceremonies and monuments honoring the contribution of such animals in Japan and other parts of Asia within a Buddhist or Shinto religion. Classes might take moments to express gratitude to animals for contributing to education, a practice that can quell mishandling of tissues and other misbehavior. Teachers using these techniques are redirecting the discomfort that students may be feeling and providing the students a more acceptable avenue for expressing emotion. This helps them become reengaged and more serious about the process. Such gestures provide some comfort to the people involved by providing an occasion to express their respect for the animals.

Dissection and Dead Animals in Education

A few studies have inquired regarding the emotional responses to having dissected animals in precollege. The use of animals in teaching elicits strong emotions at both extremes. In a study of college students looking back on their high school experiences with dissection, most reported they had performed dissection in high school, and many, but not most, experienced negative reactions (Bowd, 1993). An ethnographic study of a high school classroom dissecting guinea pigs found that most participants found the lab to be a positive experience, but some students reported it as a negative experience (Barr and Herzog, 2000).

"It might be useful to separate the elements of 'gross-out' and 'moral objection . . . or ethical revulsion' to dissection. I have students who are not grossed out at all, but who object to dissection as a practice, and vice versa . . . A 5th grade full-inclusion student, severely developmentally disabled, nonspeaking, possibly autistic, . . . loved being in science, . . . participated in every project with all his might, and no project was ever too gross, including feeling Jell-O and other materials. He . . . did every project and made every mess with us, and made a number of very astute remarks with his electronic speaking PDA. When it came to the chicken wing dissection, however, things took a different turn. We had held the ethics discussion the day before, in our preparation for the dissection. As I was briefly reminding the class of our discussion before passing out the chicken wings, I noticed this student beginning to tear up and to show signs of deep distress. His aide didn't know what to make of it . . . He was overwhelmed with compassion for the chicken, and he lacked the tools to balance that instinct with our general need to build knowledge. A lot of kids feel that, and a number of the kids knew what was going on as the boy left the room with his aide. For me that was the clearest demonstration of what kids can feel for the organism when they really think about what they're doing, and I think we err when we try to dismiss or belittle that compassion" (Bill Storm, science teacher, private communication, 2007).

LEARNING BIOLOGY

"Bodily functions, road kill, observed animal and plant reproduction, are all part of every kid's personal prior experience. It's an important access point for life science learning" (Bill Storm, science teacher, private communication, 2007). As mentioned, the most important factor influencing learning is what the learner already knows (Ausubel, 1968), and in science as in other subjects, it is crucial to start instruction from the student's point of view (Treagust, Duit, and Fraser, 1996).

As noted, children's attitudes toward science play a key role in determining their extent of interest in learning more about it, and the intermediate school years seem especially pivotal for solidifying interest (Koballa, 1995). Learning about their own bodies is a riveting interest for children, and one that is crucial for them being able in the future to manage their own medical care and understand how best to look after their own health as adults. Intermediate and secondary school are the occasions when virtually all children are still in school and can be required to complete some basic coursework pertaining to biological science and health. All children benefit from learning about their own bodies, their physiological systems, maintaining optimal health, their nutritional needs, and paying attention to how their bodies are functioning. Also, understanding through the comparative approach that other mammals have similar body systems and requirements, be it by dissection or software, and learning something about the unique sensory systems and behavior of certain animals, enhances and vitalizes the study of human biology.

Pets in the Classroom

As mentioned, many children arrive at school already having experience at home with their own pets. Through the elementary grades, children are exposed to a variety of classroom pets. The particulars vary depending on the specific interests of the teachers involved. These animals are nurtured and cared for by the children in the classroom, who watch the animal's behavior. Often these animals are there as part of the classroom context. In a few cases, the animals, especially dogs, may be brought in as visitors as a teaching tool to model appropriate behavior with animals (Craven, 1999). Many studies have documented a calming effect of the presence of a dog in a classroom (Kotrschal and Ortbauer, 2003).

Yet, as mentioned, the stereotype of children having loving companion animals is not universal in all communities. Children served by the Teaching Love and Compassion (TLC) Program of SPCA Los Angeles were found to be extremely fearful of dogs due to the prevalence of primarily fighting breeds of dogs in their neighborhoods (Zasloff, Hart, and Weiss, 2004).

Animal Knowledge Based on Dissection

The shift from caring for household and classroom pets to engaging in dissection of animals, sometimes even of otherwise pet animals such as cats or guinea pigs, inevitably is emotional. Reactions of children during and after experiencing dissection cover a broad spectrum. A study of tenth grade students in England assessed what the students had learned from dissection, and about one third reported learning nothing from the experience (Lock and Millett, 1992). Girls gave this learning-nothing response about twice as frequently as boys, and more often felt dissection was wrong. About one

Students and the Culture of Dissection

third of the students reported they had learned about body structure from the experience, a response that was more typical from boys than girls.

By now, many studies have compared traditional teaching methods involving dissection with methods involving various alternatives to dissection. The overwhelming proportion of these studies finds no difference in the learning outcomes using the contrasting methods. Only a few published papers describe comparisons of methods with intermediate and high school students (Fowler and Brosius, 1968; Kinzie, Strauss, and Foss, 1993; Strauss and Kinzie, 1994). Some unpublished work on this topic has been presented at conferences (Marszalek and Lockard, 1999; Kariuki and Paulson, 2001) or included in doctoral dissertations (Lieb, 1985; McCollum, 1987). Importantly, in a comprehensive review ten years ago, in both medical and veterinary schools, as well as in undergraduate education and high school curricula, students using interactive videodisc simulations or other high-tech simulations and lacking experience with dissections, showed equivalent knowledge when tested (Zirkel & Zirkel, 1997). Further, the simulations offer improved time efficiency (Fawver, Branch, Trentham, Robertson, and Beckett, 1990). Patronek and Rauch (2007) recently reviewed several studies involving educational alternatives, some of which pertained to dissection in precollege, confirming the effectiveness for learning when examined, even with very simple alternatives. A further point to mention is that using a computer simulation before dissection was shown to help students learn anatomy, whether or not they participated in dissection (Akpan and Andre, 2000).

As mentioned in Chapter 3, the Humane Society of the United States (HSUS: 2007c) posts and keeps updated an annotated summary of studies comparing learning outcomes with dissection as compared with dissection alternatives on their Web site. Studies are divided in this list according to whether they demonstrated an equivalent performance by students using alternative methods and traditional methods. The list includes both laboratories for dissection and physiology methods, currently comprising 35 studies. Those in which students using alternative methods performed equally to those using conventional methods comprised 18 studies; those in which students using alternative methods outperformed those using conventional methods comprised 15 studies; and those in which students who conducted dissections performed better than those who used alternative methods comprised two studies. Regardless of the ages or contexts of the students, it appears that with the most conscious and carefully designed approach, students' learning can be at least as effective with nonanimal alternatives.

Saying No to Dissection

In general, students may not realize what they have actually learned from dissection (Lock, 1994). Be that as it may, almost all feel ambivalence

(Barr & Herzog, 2000), and many are left with negative memories (Bowd, 1993). Thus, it is not surprising that some students prefer not to dissect.

For a student in a classroom where dissection is performed who wishes, for one reason or another, to avoid dissecting an animal, there is abundant information available on the Web, as well as some alternative teaching resources available to teachers by loan. While directing students to alternative resources, the teacher might also address the student's "grossed out" feelings, discuss with students the moral or ethical questions, or consider a memorial acknowledgment of the animal.

Much of the effort of organizations such as the HSUS has gone toward coaching students on choosing study methods in biology that do not involve the harming or killing of animals. Materials on the Web provide comprehensive background information on relevant legislation and access to alternatives. "Student choice in biology education: A policy guide" offers a concise overview of key questions and refers to other resources, as well as offering a consultancy service (HSUS, 2007b). Animalearn (2007) currently offers over 400 products and endeavors to offer personalized service to teachers to supply their needs for alternative resources via their loan program, and also consults with them on specific questions.

Whether by dissection or another laboratory activity, hands-on involvement in biology contrasts with reading about it. For all students, the question of dissection raises scientific and political issues and dilemmas for discussion. Children may find it meaningful to consider some of these dilemmas, even as they may be engaged in dissecting biology materials, whether these be rats or fetal pigs, or the chicken wings that are readily available in grocery stores. The inevitability of analogous dilemmas in the future speaks to the importance of early engagement in conversations about medical ethics and decision-making at an age-appropriate level.

PREPARATION FOR LIFELONG HEALTH DECISIONS

Medical care has become front-page news, sometimes almost daily. The difficulties in accessing appropriate medical care experienced by many people today are unlikely to disappear in the future. This is true even for those with excellent medical insurance coverage. We can expect that our citizens will need to make decisions as to when they require medical attention, and whether they should also see a specialist, or request an appointment with a physical therapist. Having a basic introduction to biology can facilitate a young person, later in life as an adult, in quickly ascertaining the nature of a medical problem when in conversation with a physician. An introductory level of biology can allow members of society to understand the literature on health issues and diseases that is so readily accessible on the Web. Biology might open the door to considering a future that includes entering the human health area as a professional, or as one of the many volunteers participating in health-care support.

Most young people are not concerned about health issues they may face later in life. Yet, they are exposed to family members whom they care about, and who are suffering from a variety of common ailments, ranging from Alzheimer's and Parkinson's diseases, to diabetes, heart disease, and cancer. The exposure within the family sensitizes a young person to be more attentive and interested in the topic. One can expect that a child will have greater interest in course material that pertains to an organ system that has some personal relevance due to the experience of a family member. In turn, seeing that classmates have experience with a disease can also spur more engagement in learning about a particular body system.

CHAPTER 9

The Animals Used in Teaching

In this chapter we see that in some ways learning biology is beginning to restore the emphasis on human biology that was characteristic of the early studies of anatomy with dissection, where animals were sometimes used as a surrogate but humans were preferred. With new methods of preservation, even human cadavers sometimes are available and can be used repeatedly in undergraduate education. Moreover, newer textbooks emphasize inquiry and often may include observations with data collection that students can make from their own bodies. More frequently, however, biology laboratories focus on some aspect of animal biology.

The use of animals in education for dissection is a topic of considerable controversy and strong opinions on both sides. Unfortunately, it has seemed as though much more effort is invested in offering invective against using animals than is invested in improving teaching methods and resources, although there are signs this is changing. Thus, while there is often strong opposition to using animals in dissection, we still seem to lack other resources well integrated with the teaching curriculum that offer at least equivalent learning opportunities for students.

When considering dissecting animals, questions come to mind about, how many animals are dissected, and how are they obtained. These questions, often put forward by people who primarily focus on animal welfare, are the ones that appear most often in public discussions of dissection. However, since we have little regulation pertaining to dead animals or use of animals in precollege settings, we know relatively little about the magnitude of use of animals in classrooms, and especially, how many animals of various species are dissected.

It is difficult to access balanced, basic information about dissection. Professional educators devote relatively little attention to dissection and how it may fit into the curricula and frameworks. Most of the material that is easily accessible on dissection pertains to encouraging teachers not to dissect from the standpoint of the animals involved. For example, the New England Anti-Vivisection Society, the American Anti-Vivisection Society (AAVS), and the National Anti-Vivisection Society (NAVS) all are mentioned as resources by the Urban Cat Project (2006), which addresses feline rescue, adoption, and advocacy.

The Humane Society of the United States (HSUS: 2007b) has prepared an array of materials on specific topics pertaining to dissection. These are practical and informative, and provide the sources for the information that is presented. A side banner on the dissection Web sites makes for easy navigation between the various Web pages dealing with many aspects of animals in education, including dissection, with a cost comparison of dissection versus alternatives, a booklet on student choice in biology education, and a booklet on uses of animals in higher education.

INFORMATION ON ANIMAL USE FOR TEACHING

Information available on the Web sometimes overviews the topic of dissection, but often represents a biased perspective. Depending upon the source, the experience of dissection of animals is presented as an essential part of biology with no adequate substitute, or, in contrast, the view may be offered that only alternatives to dissection should be considered. Carolina Biological (2004a), a company that supplies animal cadavers for dissection, offers a "Dissection Fact Sheet" arguing for the importance of dissection, citing a favorable quote from the National Association of Biology Teachers (NABT). They also claim that dissection helps students learn about internal organ systems and tissues, gives them an appreciation of the complexity of animals, and offers students one of the most memorable and instructive units in biology. Carolina argues that no animals sold are stolen pets and answers some questions concerning the sources of its animals, especially sharks, worms, cats, fetal pigs, and frogs. To their credit, they also point to alternatives for students who object to dissection and sell many products that are alternatives to dissection.

While the lobby against the use of live animals in teaching, science fairs, and the like is quite large, there also are recognized entities that argue against dissection of animal cadavers purchased for teaching purposes. In the 1990s, People for the Ethical Treatment of Animals (PETA: 2007) campaigned against Carolina Biological Supply, leading the U.S. Department of Agriculture (USDA) to launch a case against Carolina's animal embalming and husbandry practices. Videos taken by PETA inside the Carolina facilities, showing animals being removed from gas chambers and injected with formaldehyde without being checked for vital signs, played on the ABC network in 1990, and a complaint was filed by the USDA against Carolina charging violations of the federal Animal Welfare Act in 1991; the charges were cleared in 1994 (American Physiological Society, 1994).

The Physicians Committee for Responsible Medicine (PCRM: 2007) estimates that about half of the animals dissected in elementary and secondary schools classes in science and biology are frogs, with others typically being mice, rats, worms, cats, rabbits, fetal pigs, birds, and fish. Focusing on alternatives, PCRM presents an economic analysis, arguing that the alternatives would be more cost-effective over a three-year period than dissection of animals, and offering a comparison of various alternatives for the frog, fetal pig, cat, earthworm, or starfish, versus a package of dissecting supplies and thirty-five specimens, and a calculation of the range of savings associated with the use of alternatives.

Animalearn (2007) of AAVS announces the availability of over four hundred alternatives in an attractive print and Web-based directory, plus a loan program, and also offers a Web-cartoon sequence promoting student choice regarding dissection. Teacher, parent, and student resource centers of information are provided on the Web site. Their estimate is that about six million vertebrate animals are dissected yearly.

The most comprehensive array of documents pertaining to dissection is offered by the HSUS (2007b), by far the largest animal-advocacy organization in the United States. Documents to assist students in refusing to participate in dissection, "Student choice in biology education: A policy guide," and a review of studies assessing alternatives in comparison with dissection are among the resources offered on this site.

ANIMALS USED FOR RESEARCH, TEACHING, AND TESTING

At the outset we should emphasize that the use of animals in dissection accounts for a relatively small proportion of animals used in laboratory and scientific settings. While no exact numbers are available, one large study of almost 500 biology/science teachers endeavored to gather some indication of the species of animals used in dissection. Among the teachers who used dissection, 73 percent used frogs, 49 percent used fetal pigs, and 17.8 percent used cat dissection in their biology classes (King, Ross, Stephens, and Rowan, 2004). A few years ago in an information sheet entitled "Dissection," from Johns Hopkins University Center for Alternatives to Animal Testing (undated), it was estimated that 6 million vertebrate animals are dissected in U.S. high schools each year. Use in the United States for all educational purposes was estimated at close to 10 million vertebrates and 10 million invertebrates. Frogs, fetal pigs, and cats were listed as the most commonly dissected vertebrates. In a survey of 28 teachers in urban England, fewer than half were using vertebrate organisms or parts for dissection (Adkins and Lock, 1994). Rats and fish were the most common vertebrate organisms used. Parts of dead sheep, pigs, cattle, and rabbits were used, including hearts, kidneys, lungs, and eyes, in that order.

Little is known about the extent of animal use in precollege settings in the United States, since it is unregulated. Nor is the use of dead animals for dissection monitored. However, there are annual routine counts monitoring the use of live animals in scientific procedures that provide an indication of the number of animals used in higher education. With these figures, we can assess the proportion of live animal use that is for training and higher education, and this perhaps provides an approximate indication of trends in animal use in education in general. Since the figures in the United States exclude rodents (the vast majority of animals used), it seems more informative in this brief comment to invoke reports from England and Europe. The United Kingdom, with detailed oversight from the Home Office (2007), publishes annual reports on all procedures that use live animals. Of all procedures for research, teaching, and testing in 2005 that used animals, rats, mice, birds, and fish comprised over 97 percent. Dogs, cats, horses, and nonhuman primates, having special protection, were collectively used in less than 1 percent of procedures. Educational uses of animals, presumably only in higher education or training of professionals, dropped 41 percent over the previous year, to a total of 1,616 animals, out of a total of 2.9 million animals for all procedures at all levels. In recent years, the European Union (EU: 2007) has been striving to achieve consistent reporting practices on the number of animals used in scientific procedures among its 15 member states. Again, reflecting primarily higher education, in 2002 about 340,000 animals were used for teaching or training, including those used for dissection. This was about 3.2 percent of all procedures using animals (Sauer, Spielmann, and Rusche, 2005). In some cases, the number used in education increased due to new regulatory requirements for more extensive training of staff; this was the case in Denmark, which saw a 69 percent increase in animal use in training. The figures provided for animal use in higher education, from the EU, Canada, and the United States show that educational uses are relatively small compared with the other uses.

Discussions regarding dissection reflect perspectives representing educational and animal constituencies. The topic of dissection does not frequently emerge in the literature dealing with educational contexts, nor is it evident in libraries serving schools of education. Dissection as a precollege activity is kept visible by animal welfare organizations, which have given it considerable attention, and created documents, searching tools for teaching resources, and support and legislative avenues for students who prefer not to dissect. Finally, research studies from time to time assess the learning of students, comparing dissection with other teaching methods. While these studies do not focus on the use of animals per se, the results of the studies have implications for whether the dissection should be a preferred method, or whether another method may be equally effective.

Teachers' associations seem to have been somewhat supportive of the concept of dissection in intermediate and secondary schools, but without much discussion of how it complements the standards and goals of specific lessons, nor much consideration of the issues of using animals for this purpose.

INVOLVEMENT OF ANIMALS ASIDE FROM DISSECTION

Animals appear in classrooms in many forms and contexts, sometimes as live animals, or after dying. Body parts or even products from animals, such as feathers or claws, may be used to facilitate learning. Any educational context may feature animals on occasion, whether in actual, virtual, or symbolic form. With young children, the symbolic use is employed to motivate them to read. As an example, giraffes show up as a frequent topic in libraries and book lists, but the majority of these books are fictional accounts for your children. For young children, animals can be omnipresent in their classrooms, appearing as pets, as realistic subject matter, and as central figures in fictional accounts. Students in college or professional school, though unlikely to experience classroom pets, also may encounter animals in a range of contexts, from animal handling or husbandry courses to literature concerning animals, as well as studying parts or tissues or cellular aspects of animals.

Commonly, on college and university campuses today, students studying biology focus on cellular biology or genetics, where tissues or cells are examined rather than whole animals. In precollege early grades, teachers commonly keep classroom pets, and employ animals as a topic for presenting other material. Sometimes children are instructed in how to interact with animals, such as methods of safely meeting an unfamiliar dog. The introduction of life science as a discipline typically begins in seventh grade. In the early years of elementary school, children are exposed to live animals in their classrooms and on field trips, and encouraged to enjoy them as pets and also to learn about them. Having classroom pets, teachers can introduce some aspects of structure and function of their bodies, and notice their behavior and how it varies depending upon the behavior of others. Some teachers are able to extend these experiences by drawing on informal exposure, such as visits to museums, trips taken with the family, conversations built on reading experiences, or structured 4-H projects. In some cases, children can conduct "scientific studies" consisting of observation of animals during a zoo field trip or in observations of their pets at home or their classroom animals.

A survey of twenty-eight urban schoolteachers of biology in the United Kingdom found that live animals were more commonly used than cadavers (Adkins and Lock, 1994). Fish, protozoa, annelids, gerbils, locusts, woodlice, amphibian, and stick insects were the most frequently mentioned, in diminishing order. Most of the types of live animals were primarily used solely

Figure 9.1. Visitors get an up-close look at an African giant millipede at a public festival sponsored by Explorit Science Center in Davis, California. [Courtesy of Explorit]

for observation, although half of the teachers conducted experiments on invertebrates.

Field Trips to Zoos and Museums

Anyone visiting an active panda or elephant knows the enjoyment of getting close to real wild animals in zoos. Even seeing full-size specimens in dioramas at museums is a favored motivational shot in the arm. Teachers and parents frequently employ these trips as an adjunct for learning (Figure 9.1). Observational studies of families and classrooms in zoos and other exhibits suggest that the extent of learning and level of conversation does not really differ between families and classrooms (Tunnicliffe, 1996c). Oddly, the author questions the instructional value of these trips as part of classroom learning, since families do it as well. Nevertheless, how regularly families take such visits is being overlooked. The study does point out the perhaps missed opportunity to assign learning projects to zoo visits so that they are qualitatively different than family visits to zoos.

Science Projects or Fairs

Unfortunately, for many years, the unregulated and relatively unsupervised projects were developed by students for science fairs, sometimes

Figure 9.2. During an Explorit Science Center at Davis program presented at an elementary school, students get an opportunity to use a hand pump to inflate a pair of sheep lungs. A hose was attached to the sheep's trachea for this demonstration. [Courtesy of Explorit]

involving physiological manipulations of animals that created distress for the animals, as well as bad publicity for the poorly supervised scientific effort. With effort over the years, projects involving live-animal use, if at all questionable, are precluded from being accepted. Yet, it is still not so unusual for us to have inquiries from precollege students who are considering conducting experiments at home that involve stressful nutritional or other interventions. The lack of supervision is a problem that has not been completely corrected.

LIVE ANIMALS AND SPECIMENS USED IN INFORMAL EDUCATION

An important area of animal presentations to the public has been the development of national institutions where members of the public could view live animals or animal specimens. These facilities are repositories for live animals, often gathered from around the world, or for saving specimens gathered from dead animals to be used for viewing, learning, and even research. Over time, such venues for informal science education have multiplied and become more community-based.

Zoological collections with displays of animals, from both familiar and exotic places, are the mainstays of public institutions where people can view live animals. The high attendance at zoos by families reveals the deep fascination over time of a wide segment of the public who wish to view live wild animals. While some authorities have provided an assessment of zoos, the attraction of the animals to humans, and some features to consider when designing enclosures for animals (Hediger, 1964, 1969), others have emphasized the paradox of keeping wild animals in such unnatural settings as some zoos (Hancocks, 2001). For a historical account of zoo and aquarium history, considering ancient animal collections to modern zoological gardens, an edited volume is available (Kisling, 2001).

Zoos welcome visitors, including classrooms on field trips as well as families. In most zoos, visitors view the animals in day quarters that are constructed to provide a realistic appearing natural setting for the animals, and at night many of the animals are moved into smaller quarters where handling and feeding are simplified. The zoos generally continue breeding at least some of their animals, with advocacy for breeding as a strategy to protect endangered or threatened species, plus they may sell or trade surplus animals to other facilities as a redistribution method.

In contrast to zoos, sanctuaries house animals that have become "homeless" due to injury or irresponsible caregivers, and these facilities do not allow further breeding of their animals because their primary goal is simply to offer a comfortable refuge. Visitation to such facilities is often discouraged and sometimes limited to a few days a year as a special occasion. Performing Animal Welfare Society (PAWS: 2007) is an example of a sanctuary that features somewhat conventional cages for bears, large cats, some monkeys, and other mammals at its facility in Galt, California. They also manage a large group of neutered hoofstock in a large enclosure at the Amanda Blake Memorial Refuge and Museum. Also as part of PAWS, expansive enclosures at ARK 2000 (a 2300-acre natural habitat, captive wildlife sanctuary) currently house tigers, African elephants, and Asian elephants. Performing Animal Welfare Society offers a few events each year for the public to visit the Galt Sanctuary or the ARK 2000 Sanctuary. Sanctuaries are not limited to exotic wild animals. Animal Place (2007) in Vacaville, California, provides homes for pigs, cows, chickens, turkeys, goats, sheep, and dogs, and offers a virtual tour for people who wish to visit the animals. During summer months, Animal Place offers a couple Sanctuary Farm Tours each month as well as volunteer opportunities with an orientation training every few weeks for prospective volunteers. Tours are specialized, with some focusing on children and others on seniors, and are offered at modest cost. Student internships are available for college credit.

People involved with both zoos and sanctuaries engage in advocacy reflecting their separate visions of how best to serve their animal constituencies. Whether to allow trade in animals for profit, whether importation of animals should be permitted, the extent of visitation to animals, views on the animals' involvement in educating the public, and the types of legislation that are supported are some of the issues on which groups may differ.

Another type of informal education involving animals are projects such as those run by Guide Dogs for the Blind, in which participants raise dogs for a very worthy cause. An animal activity, more-or-less fading away, is 4-H, which personifies informal education regarding animals. Children are assisted in raising and showing animals. Along with cows, goats, horses, pigs, and chickens, rabbits, dogs, and cats are among the young animals adopted by young people and cared for as they grow up.

Finally, there is the increasingly popular opportunity for children to be involved in the care and training of animals in shelters and humane societies. The goal is to eventually solve problems of homeless cats and dogs. These facilities are nearby in almost every community, providing opportunities for children to volunteer and gain experience with animals.

DISSECTION AND USE OF SPECIMENS
IN PRECOLLEGE CLASSROOMS

Focusing now on the process of dissection or the use of specimens in classrooms, we notice that these are familiar aspects of high school biology, and, while somewhat costly, it is relatively convenient to get equipped.

Acquisition of Specimens

Specimens for dissection and study can easily be ordered on-line from a supplier such as Carolina Biological Supply Company (2007), WARD'S Natural Science (2007), or NASCO Online Catalogs (2007). The catalogs are numerous and distributed frequently. As alternative products have become popular, the suppliers have nimbly incorporated them into their catalogs alongside the preserved specimens. Discount prices are offered by WARD'S for ordering 100 or more frogs, rabbits, cats, or guinea pigs, whereas NASCO features a pail of 25 rats or package of 10 cats.

Among whole organisms, rabbits are embalmed, and under the triple injection option can be prepared with the arterial system injected with red latex and the systemic veins injected with blue latex, and even the hepatic system injected with yellow latex. Fetal pigs are available, and the red and blue injection is an option. Cats are available with the same preparation methods, including the three colors of injection. Cats that have been skinned can also be ordered, saving dissection time and reducing the exposure to formalin. Another option is that the cat be pregnant and injected with color-coded latex, or not.

Among organs, sheep brains are an example, with a variety of options for preparation and packaging. The dura mater can be removed with the optic chiasma intact; or the dura mater can be intact with the hypophysis and cranial nerve roots intact. The arteries can be injected with red latex and veins with blue latex, or not. A sheep brain prepared as a prosection with numerical labeling of the cranial nerves and other vital parts of the brain is also available, with a text for the labels. Sheep lungs that are air-dried and inflated, and then spray-coated with clear plastic are available, as a whole lung or a section. Cow specimens are available, such as eyes, hearts, spinal cords (4″ portion), and bull testicles.

It seems that virtually any animal specimen imaginable is available for purchase. Especially with the World Wide Web, marketing of specimens is efficient and targeted to special niche markets, with a variety of dissection kits. One Web site is directed toward home school science and carries many specimens and dissecting kits suitable for a single person, at modest cost (Home Training Tools, 2007). Animal hair, chameleons, cockroaches, and amphibians—an array of specimens is available from WARD's Natural Science (2007). Einstein's Emporium (2006) emphasizes exotic specimens ranging from zebra skins to butterflies and moths to spiders and scorpions, mounted snakes, and seashells and mollusks.

Suppliers

An incentive for dealing with major providers of specimens is that they offer uncomplicated, one-stop shopping for dissection specimens, as well as curricula or science programs, and some live plants and animals. Carolina Biological supplies Science & Technology for Children for grades K-6, and Science and Technology Concepts for Middle Schools (2004b). Both of these were developed by the National Science Resources Center, operated by the Smithsonian Institution and the National Academy of Sciences. Live animals offered include vertebrates such as Dwarf African frogs and fish, as well as various arthropods, microorganisms, and other invertebrates, along with a Web-free guide on care and handling of living materials (Carolina Biological Supply Company, 2004c).

Frequently concern arises regarding how or where animals are obtained for dissection. This concern may often be overstated, and it can be difficult to track down the extent to which the sources are not appropriate. In this light, critics sometimes focus primarily on a particular species that may be dissected. Advocates for cats maintain a Web story from 1994, purportedly from the World Society for the Protection of Animals (WSPA: Messybeast.com Cat Resource Archive, 2007), reporting on an investigation of a trade in slaughtered cats from Mexico in which the cats were prepared as preserved biological specimens, and then exported to the

United States. Although this story is still posted on the Web, it does not appear on the WSPA site, and it is difficult to know whether the story is authentic. The Urban Cat Project (2006) lists classroom dissection of cats as a concern, saying that over 100,000 cats are dissected annually in U.S. high schools. Their site refers to class B dealers obtaining cats by stealing stray pets, and mentions that the dealers are notorious for animal cruelty.

Costs

One practical approach is to examine the cost of dissection. The particular species of animal or organs to be dissected is a first decision. Cats, fetal pigs, frogs, and earthworms are commonly considered. The considerable cost of the fixed specimens often is the determining factor in choosing the activity, since budgets usually are limited. All of these species are available from commercial suppliers, along with the dissecting tools, pans, and gloves required to set up the laboratory.

Focusing on specific organs or body systems is another approach teachers may use. They may decide to purchase formalin-fixed calf hearts or brains for dissection. Alternatively, they may arrange to obtain fresh specimens from a meat market or slaughterhouse. Or, just to provide a view of bones and joints, some teachers purchase chicken wings and supervise a laboratory based on that dissection. Sometimes these laboratories are supplemented with educational models or charts, or computer programs that enhance and extend what can be seen in a single dissection.

The toxicity of formalin is well known (Fox and Benton, 1987), and dealing with formalin specimens is also an unpleasant experience, as well as having problems of flammability, and causing of allergies, burns, and possibly cancer. Dissection in the presence of formalin requires good ventilation, as mentioned in the National Science Teachers Association (NSTA) guidelines. Methods for washing the specimens can reduce the extent of distribution of noxious fumes (Rosenberg and Fitch, 1998). They allude to commercial suppliers using a nontoxic holding/shipping fluid that is not identified, and recommend a 2 percent aqueous solution of ethylene glycol for holding specimens during a one-semester course. Working with the concentrate requires caution, but the 2 percent solution is recommended as nonirritating to students. This holding solution prevents fungal growth and keeps tissues pliable, without extracting much formalin from fixed specimens. Newly arrived sharks can also be preserved in this solution. As students do their dissections, formaldehyde in the specimens must be neutralized, for example with a 20 percent solution of InfutraceTM (Rosenberg and Fitch, 1998). In dissecting cats, the authors recommend discarding the skin and shipping bag. Apparel should include gloves, lab coats, and goggles, without contact lenses. Distributors endeavor to reduce the toxicity of specimens. Those provided by Carolina Biologicals are packed in "odorless *Carosafe.*" WARD'S Natural Science (2007) lists formaldehyde-free preserved specimens.

Of course, alternatives may provide a way to reduce costs. Two organizations have worked on this question, information of great relevance to administrators or department heads who oversee budget planning for their science classrooms. In worksheet figures developed by HSUS (2007b), it is assumed that dissected animals are obtained from Carolina Biological Supply and calculated how expenses compare over a three-year period for purchasing alternatives versus animals (cat, fetal pig, or bullfrog) to dissect.

The economic analyses such as those by the HSUS (2007b), or the PCRM (2007), profile that alternatives to dissection have a high cost at the beginning but tend to be reusable, whereas the specimens used for dissection are consumed and new ones need to be purchased each time the laboratory is offered. Thus, once the alternatives are purchased and the laboratory is equipped, future laboratories have lower expenses even as the costs of dissecting materials become more expensive over time. Browsing recent catalogs of laboratory supply companies shows rats cost around $5 each and fetal pigs around $10. With students working in pairs, a single rat dissection for 160 students would cost $400.

Frogs, fetal pigs, and cats are the vertebrates most commonly dissected, as well as dogfish sharks, perch, rats, pigeons, salamanders, rabbits, mice, turtles, snakes, mink, foxes, and bats. Invertebrates include crayfish, grasshoppers, earthworms, clams, sea stars, squid, sea urchins, and cockroaches. Among those animals taken from the wild for dissection are frogs, dogfish sharks, salamanders, birds, snakes, turtles, fish, and most invertebrates (HSUS, 2007b). Carolina Biological (2004c) points out that many sharks and worms are already dead when purchased from fishermen and the fishing bait industry.

Plastination

Finally, a way exists to have your actual specimens and not consume them. The preparation of odor-free plastinated specimens that are easily visualized has revolutionized the study of anatomy. In this process invented and patented by Gunther von Hagens (BODY WORLDS, 2007b), the water and fat in tissues are replaced with a polymer. With this invention, suddenly it became possible to present any type of human or animal specimen esthetically, offering biological knowledge for viewing by people of all ages.

With the advent of plastination of animal specimens, it has become possible for students to view and examine real organic material without exposure to formalin and decomposing tissue. Many institutions have found that purchased plastinated specimens are not completely satisfactory, and have initiated their own plastination processing using the patented process of Gunther von Hagens, including the University of North Carolina at

Asheville and the University of Tennessee (Stuart and Henry, 2002) as well as the University of California, Davis. The dry, odorless specimens are also reusable, and can be prosected by a professional to expose the most essential parts with appropriate labels. Multiple laboratory stations can present various aspects of the same organ in a different prosection.

Since precollege institutions are not equipped to implement preservation technology, partnerships between universities and school districts can facilitate delivering this technology to the public schools. For example, the University of Michigan Plastination Laboratory and the Dow Corning Corporation collaborated with a local school district (Douglass and Glover, 2003). The Bayer Corporation partnered with the University of California, Davis, in a model project delivering plastinated hearts and *Virtual Heart* software to sixteen biology classrooms in the Sacramento Valley (Zasloff and Hart, 1997).

Virtual Bodies: The Computer Age of Dissection

New computer technology has transformed the possibilities for providing for effective and efficient learning of human and animal biology in the absence of old-fashioned dissection. The new technology has been taken up especially in veterinary medical education, where new curricula reflect capabilities of computer software, and these are largely replacing dissection (Hart, Wood, & Weng, 2005). Unlike formalin-preserved or even plastinated specimens, realistic three-dimensional (3D) images on software provide virtual dissections of fresh bodies and are annotated with pop-up labels and texts. This revolutionary technology is not yet available in precollege biology classrooms. However, offering virtual organs in 3D should improve performance over prior videodiscs. The range of resources available for computer software is a topic of the next chapter.

CHAPTER 10

New Teaching Resources in the Computer Age

Although mention of dissection today is largely absent in curricular materials for biological science and presentations of educational standards, the practice of dissection continues as a tradition that is part of the precollege culture of teaching biology. The practice of dissection is not subject to any census count of numbers of animals or overall review. Thus, estimates of the number of animals dissected nationally in precollege classrooms are approximations, and not based on any monitoring. Where scrutiny has arisen concerning dissection in precollege classes, it has emphasized communication with students to provide them some supportive protection against being required to perform dissection. Some states have implemented legislation requiring that students have the opportunity to opt out of dissection if they prefer.

Critics of dissection, typically focusing on animal welfare, often emphasize that teachers should stop using dissection as a method of teaching biology. But little attention is directed on what might supplant dissection. From conversations with teachers, we hear that dissection carries appeal in generating some excitement and involvement. The motivational effects of dissection are palpable to teachers, who every year witness the excitement of at least some students associated with dissection. If the dissection is stopped, a major question concerns what will supplant it as a landmark experience for students. Perhaps it would be feasible to generate similar excitement if high-end computer technology were harnessed to teach biology. If one were to seek some resources to supplant dissection, how would they be produced, by whom, and how would the effort be supported?

With a lack of oversight and funding directed toward dissection in pre-college education, major resources have not been available to develop improved teaching materials, nor has there been an emphasis on that need. The lack of discussion of dissection within the context of science educational standards, the lack of financial and technical support for alternative teaching resources, and the lack of stimulating informative teaching materials for teachers are three substantial barriers that work against mainstreaming alternative teaching resources in precollege education (Hart, Wood, and Weng, 2006). In the current professional literature, educational materials and research papers have largely ignored dissection and teaching methods, and have focused instead on issues such as equal access to quality instruction and diversity.

In contrast to dissection in the precollege years, uses of animals in dissection and surgery in the professional medical and veterinary schools have been steadily criticized and scrutinized. This is where alternative teaching materials and methods have been developed, as professional talent and financial resources have been directed toward methodically shifting away from consumptive uses of animals. In veterinary medical education, a powerful coalition of insiders developed for changing animal use and developing alternatives, including veterinary students lobbying for change, technicians and professors creating new teaching resources, and administrators providing funding to support creation of the new resources (Hart and Wood, 2004). Over three decades, new teaching alternatives have materialized with contributions from virtually all veterinary schools and in this process transformed teaching methods (Hart, Wood, and Weng, 2005).

RESOURCES FOR TEACHERS

Teachers are provided textbooks that often come with ancillary materials that can be useful for laboratories. In addition to textbook companies, many other organizations are engaged in creating new biology resources or learning environments and making them available. Some of these and their Web addresses are listed in Table 10.1. The basic repertoire of teaching resources for classrooms includes human specimens, reusable animal specimens, simulations, imaging, software, videos, charts, and models.

Human versus Animal Specimens

Most instruction in animal biology has an ultimate purpose to teach about human anatomy and physiology. The use of animals as a surrogate for humans has a long history, but modern methods of tissue preparation now make it feasible to present real human exhibits. Further, some institutions have human cadavers available through donor programs, providing opportunities even for undergraduates to study anatomy, including human specimens in the laboratories. Some undergraduate courses provide laboratory experience in working with prosected body parts from donated specimens. These prosections are professionally dissected parts of hands and feet, legs and arms, and various organ systems. Those students planning to go into dentistry, optometry, nursing, physical therapy, occupational therapy, or anthropology are attracted to these courses. UC Davis is among the few universities in the United States offering this type of program to undergraduates. To honor these donors, the campus established a grove and bench dedicated to body donors.

Study of animal anatomy to learn human anatomy is a more typical approach. Prosections and plastinated specimens can become the mainstay of some laboratories, and be reused year after year. Acquiring the specimens initially is a challenge, but once they are obtained and suitably prepared, they can be used repeatedly, and stored in drawers when not in use.

Software, Videos, Charts, Models

While virtual simulations and new types of imaging are revolutionizing biological education by providing high quality visual displays, conventional software, videos, charts, and models are still available and provide the basic pictorial and three-dimensional views of animals and organ systems. Interactive frog dissection software has been available for over a decade (Sweitzer, 1996), but the two programs reviewed were seen to fall far short of dissecting a real frog in terms of realism and material covered.

HIGHER EDUCATION

A brief discussion of the evolution of uses of animals in higher education may be instructive in providing further guidance for the change in use of animals in the precollege situation. The content, curricula, and coursework used in scientific higher education are typically created, monitored, and reviewed by the faculty who are engaged in teaching these courses. Universities, colleges, and professional schools would not exist without the students who volunteer to attend and become educated. Thus, teaching is a preeminent priority of faculty, and their performance in the classroom is reviewed as part of their assessment for merits and promotions. Faculty members are responsible for a varying number of lectures each year, depending on the institution and particular teaching assignment. Faculty assume a central responsibility in charting their courses, updating their material, writing syllabi and textbooks, planning laboratories, and revising course outlines as the subject matter and methods evolve. An interesting contrast is that the teaching responsibility of higher education faculty is considerably less than of precollege faculty with regard to the number of hours of lecture and laboratory. Higher education faculty have time out of the classroom and resources available for developing new course materials. They assume a responsibility for designing their courses, and they

Table 10.1 Teaching Resources

Title and URL	Producer	Scope
ASCD Association for Supervision and Curriculum Development http://www.ascd.org	Association for Supervision and Curriculum Development	ASCD resources are intended to offer effective techniques for teaching in order to improve student achievement.
ASTC Association of Science-Technology Centers http://www.astc.org/pubs/index.htm	Association of Science-Technology Centers	An association of science centers and museums, ASTC is dedicated to furthering the public understanding of science among increasingly diverse audiences. ASTC encourages excellence and innovation in informal science learning, providing information via books, periodicals, and other online resources. It provides professional development for the science center field, promotes best practices, supports effective communication, strengthens the position of science centers within the community at large, and fosters the creation of successful partnerships and collaborations.
CALF http://www.calf.vetmed.ucdavis.edu/calf.html http://169.237.113.35/FMPro?-db=Products.FP5&-lay=CGI&-token=%5BFMP-currenttoken%5D&-format=search.htm&-view	Computer Assisted Learning Facility, School of Veterinary Medicine, University of California, Davis	Products developed and produced by the School of Veterinary Medicine at UC Davis specifically for veterinary medicine education include CD-ROMs (such as The Virtual Heart), videos, and vascular access models.
CSTA California Science Teachers Association http://www.cascience.org/index.html	California Science Teachers Association	CSTA organizes conferences and programs, supports professional publications, and provides timely access to relevant legislation, news, and information.

(*continued*)

Table 10.1 (*continued*)

Title and URL	Producer	Scope
California Selected State Resources http://www.ed.gov/about/contacts/state/ca.html	US Department of Education	Selected state resources for California, including contacts, accountability, and statistics.
State Contacts and Information: Selected State Resources http://www.ed.gov/about/contacts/state/index.html?src=ln		Selected state resources for each and every state, including contacts, accountability, and statistics.
Center for Mathematics and Science Education, MASE Center http://www.csus.edu/mase/	MASE Center at California State University, Sacramento	The mission of the MASE Center at California State University, Sacramento, is to improve the quality of science teaching and learning in the public and private schools of the Sacramento region. Two areas of particular importance are the professional development of experienced teachers and development of preservice teachers.
FOSS - Full Option Science System http://www.lhs.berkeley.edu/foss/index.html	Lawrence Hall of Science, University of California	FOSS is a research-based science program for grades K–8 developed at LHS with support from the National Science Foundation and published by Delta Education. It is designed to provide enrichment for students and support for teachers and administrators, and is dedicated to improving the learning and teaching of science.
Library of Congress http://www.loc.gov	Library of Congress	More than 10 million primary sources online, including lesson plans, activities, and educational resources.

(*continued*)

Table 10.1 (*continued*)

Title and URL	Producer	Scope
For teachers http://www.loc.gov/teachers/		
NABT National Association of Biology Teachers http://www.nabt.org/ NABT Position Statements http://www.nabt.org/sites/S1/index.php?p=26	National Association of Biology Teachers	The National Association of Biology Teachers, NABT is the primary professional association for those in life science education. Resources and materials useful to teaching include Biology Teaching Preparation Standards for Middle & Secondary Teachers, Ethics Statement for Biology Teachers, NABT's Statement on Teaching Evolution, The Role of Biology Education in Addressing HIV & AIDS, Role of Laboratory and Field Instruction in Biology Education, and The Use of Animals in Biology Education.
NAHEE http://www.nahee.org/	National Association for Humane and Environmental Education, Humane Society of the United States	NAHEE is the youth education division of the HSUS. The Web site, Humane Society: Youth serves as the HSUS youth education affiliate. Included are efforts to provide teaching materials, professional development, and other support to teachers and humane educators.
NCISE National Center for Improving Science Education http://www.wested.org/cs/we/view/pj/445	West Ed	National Center for Improving Science Education (NCISE) works to improve science and mathematics education for K-12 students. Included in this is preparing science and mathematics teachers for the classroom. NCISE partnered

(*continued*)

Table 10.1 (*continued*)

Title and URL	Producer	Scope
		with the Council of Chief State School Officers (CCSSO) to develop the Science Framework and Specifications for the 2009 National Assessment of Educational Progress (NAEP).
NSRC National Science Resources Center http://www.nsrconline.org/	Smithsonian Institution and National Academies	The Smithsonian Institution and the National Academies jointly established the National Science Resources Center (NSRC), with the mission to improve the learning and teaching of science for all students in the United States and throughout the world. The NSRC offers integrated leadership development for school districts, professional development for teachers, science curriculum for K–8 students, and internships for high school and college students.
NSTA National Science Teachers Association http://www.nsta.org/	National Science Teachers Association	NSTA works to provide and expand professional development to support standards-based science education, while also supporting innovation in science teaching and learning, curriculum and instruction development, and improved assessment.
Official Positions http://www.nsta.org/about/positions.aspx		Positions include Responsible Use of Live Animals and Dissection in the Science Classroom, Science Competition,

(*continued*)

Table 10.1 (*continued*)

Title and URL	Producer	Scope
		Elementary School Science, and National Science Education Standards.
Project 2061, AAAS http://www.project2061. org/	American Association for the Advancement of Science	Project 2061 is an AAAS initiative to advance literacy in Science, Mathematics, and Technology. The project includes suggested ideas, lesson plans, curricula, and courses for alternatives to dissection.
Atlas of Science Literacy http://www.project2061. org/publications/atlas/ default.htm		AAAS Project 2061 and the National Science Teachers Association have released the two- volume Atlas of Science Literacy, a collection of conceptual strand maps that show how students' understanding of the ideas and skills that lead to literacy in science, mathematics, and technology might develop from kindergarten through 12th grade.
SEPUP Science Education for Public Understanding Program http://www.lhs.berkeley. edu/sepup/	Lawrence Hall of Science, University of California	SEPUP (The Science Education for Public Understanding Program) creates innovative science curriculum for use in grades 6–12 education, engaging students and teachers in issue-oriented science. Located at the Lawrence Hall of Science at the University of California at Berkeley, SEPUP supports excellence in science education by developing research-based curricular materials and providing teacher professional development.

(*continued*)

Table 10.1 (*continued*)

Title and URL	Producer	Scope
SEP Science and Health Education Partnership http://biochemistry. ucsf.edu/~sep/	UCSF and SFUSD	The Science & Health Education Partnership (SEP) is a collaboration between the University of California, San Francisco, and the San Francisco Unified School District. Scientists and educators from both organizations work in partnership to support quality science education for K-12 students.
Resources http://biochemistry. ucsf.edu/%7Esep/ resources.html		
Science Service http://www.sciserv. org/index.htm	Science Service	Science Service's mission is to advance public understanding and appreciation of science through publications, educational programs, science fairs, and scholarship competitions. Science Service administers two prestigious competitions in precollege science, encouraging students to utilize and strengthen their knowledge in science, math, and engineering.
Society for Neuroscience: Information for Educators http://www.sfn.org/ index.cfm?pagename= InformationForEducators	Society for Neuroscience, Washington DC	SfN is committed to partnering with educators to engage students in learning about the brain and nervous system. By integrating current, accurate neuroscience content into science teaching, SfN will assist in building capacity for K-12 teachers and enhancing science education.

(*continued*)

Table 10.1 (*continued*)

Title and URL	Producer	Scope
Valley Oak Science http://www.djusd.k12.ca.us/valleyoak/bstorm/ http://www.djusd.k12.ca.us/valleyoak/bstorm/biosafarishome.htm	Bill Storm, Valley Oak Elementary, Davis, CA	Valley Oak Science serves 4th through 6th graders in two classrooms dedicated to science instruction. Instruction is standards-based, with curriculum delivered through the Investigation and Experimentation strand of CA State Standards.
WestEd http://www.wested.org/	WestEd	WestEd is a research, development, and service agency, intended to enhance and increase education and human development within schools, families, and communities.
When science meets the public http://archives.aaas.org/publications.php?pub_id=108	AAAS publication; 92-06S American Association for the Advancement of Science	The proceedings of a workshop organized by the American Association for the Advancement of Science, Committee on Public Understanding of Science and Technology, February 17, 1991, Washington, DC.

can recruit assistance from technology support and other departments on campus to help them in continuing to keep their courses fresh and reflecting current material. As faculty in higher education settings revise their courses, the new course outlines are submitted for approval to the appropriate curriculum committees within their department, school, and campus. Thus, the instructional content and material are conceived, designed, and developed by those who do the teaching. This extensive engagement in teaching that is characteristic in higher education would not be feasible in precollege education, where teachers have much less discretionary time, but making room for some activities along this line would be a worthy goal of administrators. Historically, the most consumptive educational use of animals per student traditionally occurred with veterinary students learning basic skills of the profession in courses such as anatomy, physiology, and surgery. Veterinary students must become proficient in treating animals in just four years of veterinary school, sufficient to provide excellent care to clients' animals. Veterinary students face an imperative to become knowledgeable and skilled in working with animals beyond that required for students in any other discipline.

The strongest initiatives to create improved methods that are less consumptive in using animals have come from veterinary schools, where the use of animals in education has been the greatest. What we have seen in the veterinary community is a gradual process of transitioning over to alternatives, so that by now, alternatives have been mainstreamed and have largely replaced consumptive uses of animals.

To lay out a framework of a vision for what might be done for teaching of precollege biology, it is useful to review the process by which the introduction of alternatives to the use of animals in veterinary medicine has transformed much of the preclinical instruction. To begin with, faculty, staff, students, and administrators working together over a couple of decades have created a cultural and curricular change within veterinary medicine (Hart and Wood, 2004). This has involved a significant allocation of financial resources, as well as commitment of creativity and talent toward developing new strategies for creating laboratory materials. Faculty members wrestle with producing effective instructional laboratories each year. Coming from a knowledgeable position concerning what students need to learn, faculty have shown great ingenuity in developing new approaches such as a mechanical model that illustrates the glomerular filtration function of the kidney and a "bone box," modeling a broken bone in a foam lined box on which students could learn skills of bone-plating, a skill requiring coordination of two pairs of hands in a small space. The plastination of various organs has already been mentioned in Chapter 9 and anatomy laboratories have gradually shifted from using dissection to using various plastinated specimens, supplemented by software models and other materials. Animal models allow students to practice collection of blood samples, place catheters within veins, pass a tube within the trachea, and apply bandaging.

In the realm of teaching anatomy, the advent of computer software has led to many products for veterinary education, including the CD-ROM, *The Virtual Heart*. This product included a version developed in collaboration with high school teachers and is available for purchase (UC Davis School of Veterinary Medicine, 2007). It has proven popular among precollege teachers.

The innovative progress of the UC Davis School of Veterinary Medicine has led to thinking about possible applications beyond the school and a course was organized for undergraduates with ten laboratories, each organized on one body system comprising about thirty laboratory stations with reusable materials. In theory, and with sufficient startup funds, such a project could be modified for precollege science biology instruction.

Also at UC Davis, a project team is developing a prototype that adapts images from "The Virtual Human" to a three-dimensional interactive program accessible on personal computers. This is part of a larger goal

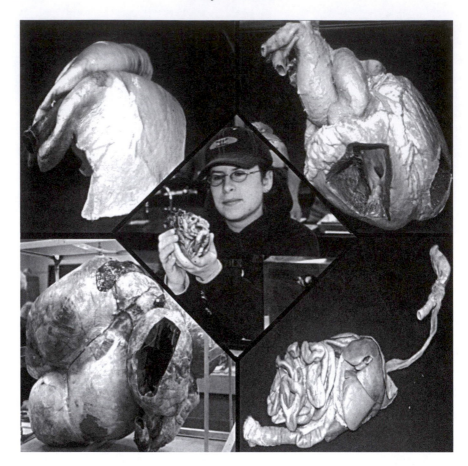

Figure 10.1. Students can make side-by-side comparisons and see the extreme size contrasts among mammals in their lungs, digestive tracts or stomachs, and hearts in laboratories with plastinated or freeze-dried organs. The structure of the liver and gall bladders, with their complex circulatory system, are better depicted by a model than a preserved organ. These images are from the University of California, Davis, undergraduate course.

to cover the entire body with engaging software of three-dimensional virtual images and is referred to as *BioSafaris*. In addition to the software programmers and anatomy professors on the project team, teachers are engaged and provide feedback as the project evolves.

In medical schools, animal cadavers and live animals were previously used as surrogates for humans in teaching medical students and conducting physiology experiments. Although medical education has not spearheaded animal alternatives for teaching to the same extent as in veterinary medical education, a variety of human simulations have been developed that supplant procedures that formerly may have involved the use of animals. Virtual reality simulators and interactive human manikins are now used

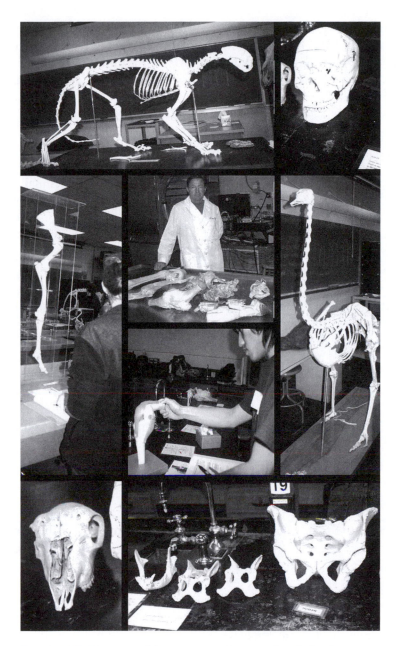

Figure 10.2. During a laboratory session on the skeletal system, students in an undergraduate comparative anatomy course may be surrounded by full-sized skeletons of various species, comparative mountings of legs or skulls, and a mounted knee joint, like these at the University of California, Davis. Most of the laboratory stations feature brief explanatory text to help guide the students' investigation of the materials. Instructors who rotate among the students are available to answer questions.

for some of the training of future physicians and nurses in procedures such as endoscopy, laparoscopy, anesthesia, trauma management, angiography, and needle insertions (Balcombe, 2004). The simulators which are still costly are not yet widespread and their use has not yet supplanted these uses of animals for training.

PRECOLLEGE EDUCATION

Intermediate and secondary school science teachers have compelling responsibilities and lack the discretionary latitude and the time to develop their own teaching materials. Typically, they have multiple course preparations each day, a large number of classes and students, tight scheduling constraints, and limitations in their technological and laboratory capabilities. In about twenty-two of the fifty states in the United States, teachers are expected to rely primarily on teaching materials that are adopted by their state or local boards of education, and then supplement these with appropriate laboratories and activities as feasible.

Many states mandate which textbooks their schools can adopt in an effort to ensure alignment between curricula, teaching and state testing. Critics have singled out California as being particularly specific in defining the forms of presentation, pedagogy, and technology that are acceptable (Muzslay, 2006). We point to California here as an example of how the requirements and budget for selection of materials can add layers of complexity to the constraints facing teachers in the classroom. In California, state instructional materials moneys are tied to the list of materials that have been adopted by the California State Board of Education (K-12 Alliance, 2006). For the 2000 adoption criteria, the Science Framework Committee and the State Board of Education developed five criteria: content, program organization, assessment, universal access, and instructional planning and support. Although normally 70 percent of the Instructional Materials Funds (approximately $30 per student in 2000) are to be allocated for the purchase of state adopted materials, a petition process is available to obtain a waiver, to allow school districts to spend their Instructional Materials Fund on resources that are not state adopted, but better meet the curricular goals of the district.

The entire process in California for school districts purchasing instructional materials, as described above, already seems fairly complicated. However, the specifics for this process are revised each year in the California Budget Bill and are subject to political pressure. The pertinent California State Statutes and the budget available are subject to annual revision, creating a moving target that districts must continually review.

Similar processes are at work in other states and local districts where constraints are imposed. These constraints together, in the aggregate, are major forces affecting the acceptable content of textbooks. Some of the larger states are particularly influential in this regard, as they have developed their own standards to be followed in addition to the national standards, and created their own list of adopted materials. With their adoptions of certain textbooks, Florida, Texas, and California exert a great impact on other states. Concerns have been raised about the entire process of textbook adoption, as being too influenced by politics, limiting the updating of materials and seeking a lower level rather than one reflecting the individuality of districts (Ezarik, 2006). Surprisingly, it also is asserted that most of the top-performing states do not adopt specific textbooks.

This should make it clear that individual science teachers receive clear directives concerning appropriate subject matter and materials to teach and use. A further concern is that their students are administered an array of tests to measure their learning and teachers are expected to teach effectively so as to assure reasonable performance from their students.

What seems to be missing as a major concern in the current literature and resources, though, is an emphasis on the science laboratories for students that could provide them excitement and bring life to the scientific material, illustrating how questions are answered, new knowledge is generated, and problems are solved with scientific methods.

TEACHERS' PROFESSIONAL ASSOCIATIONS

Professional organizations are a valuable resource available to teachers. These organizations may set policies regarding uses of animals and also provide information on teaching materials and conferences to be held with presentations pertaining to topics of interest. The National Science Teachers Association (NSTA : 2007) is an example, which offers a resource, NSTA's Suppliers Guide. This lists an abundance of company listings, including equipment/supplies, media producers, computer hardware/software, textbook publishers, resource materials publishers, and tradebook publishers. Elsewhere on its Web site, NSTA also provides reviews of outstanding science tradebooks, and news items. Teachers perusing this Web site on a regular basis would likely see mention of resources of possible interest.

The National Association of Biology Teachers (NABT : 2007) is another major resource where teachers can learn about and access resources. Like NSTA, NABT sponsors excellent publications and conferences. Much of the discussion emphasized on the Web site pertains to teaching evolution and intelligent design, an ongoing controversial topic for which the organization has issued a specific statement (National Association of Biology Teachers, 2003). Although dissection is not profiled, NABT provides a recent position statement on the use of animals in biology education (National Association of Biology Teachers, 2003). NABT states that "no alternative can substitute for the actual experience of dissection or other use of animals" and urges teachers to "be aware of the limitations of alternatives." It goes on to suggest that when the teacher determines that the objectives can be met effectively without dissection, alternatives may be used such as models and various forms of multimedia. NABT encourages teachers to be sensitive to substantive student objections to dissection and provide appropriate lessons for those students where necessary.

The NABT statement also refers to a more detailed set of ten principles prepared by the Institute of Laboratory Animal Resources (ILAR : 2006). These guidelines pertain especially to the use of life animals, either as classroom pets, or as involved in experimental studies. It includes a recommendation that if experiments are being done in precollege classrooms, a panel can be appointed by the school principal that includes a science teacher, a teacher of a nonscience subject, and a scientists or veterinarian with expertise in the subject matter involved. Such an oversight committee is not suggested for dissection, but it seems as though this would be appropriate.

The California Science Teachers Association (2007) provides a Web-based list of science education suppliers that are categorized as to textbooks, Web sites, equipment, workshops, fieldtrips, life science, teacher resources, student resources, and free resources. They provide ongoing monitoring of legislative initiatives that are underway and offer lively conferences. Currently, dissection seems not to be profiled by these associations, except for some articles appearing in their journals.

SUPPLIERS OF TEACHING RESOURCES

As described above, the acquisition of textbooks is often a complex process, even a political decision affected by a particular portion of the content of the materials and specifications about how funds can be used. For many, or even most, teachers, the textbooks are purchased without their input or comment concerning the reading level or approach that might be most effective with particular groups of students.

On the other hand, planning laboratories seems to be quite a different process, since it is not subject to the extensive review process in place for textbook adoption, nor is there an effective support system for managing the production of laboratories that would effectively complement the lessons that teachers are seeking to employ. The teachers are really on their own to acquire laboratory materials and figure out something that would be interesting and instructional, but typically they have little budget to use for this purpose. Laboratory materials tend to be consumable, but the teachers can acquire some models, charts, and other resources that are reusable, and over time, they collect a cupboard of materials that enhance the laboratories they offer.

Textbook Companies

The technology for instruction at the college level does not necessarily spill over into the precollege environment. High school and junior high school classes are affected by the textbook adoption process of their state, and the budgetary processes of their schools. Biology textbooks are reviewed, and adoptions are subject to political pressures related to the content. A further deficit today is that the array of textbooks available on the market is not readily available in libraries. In many cases, curricular libraries have been dismantled in universities. The materials are expensive, so teachers cannot conveniently review and acquire promising materials that have not been purchased in their district.

Laboratory Suppliers

With regard to science laboratories, perhaps the most readily visible and accessible resources are those offered by commercial suppliers. Carolina Biological Supply Company and NASCO (2007) are two examples of companies that cater to science teachers, producing catalogs frequently and comprehensive Web sites. Featured by the California Science Teachers Association, NASCO offers "the science teacher's favorite catalog" with over 65,000 unique products to meet the needs of teachers, including sections on anatomy, physiology, dissection, entomology, live materials, and preserved materials, including various kits, skeletons, models. Similar resources are available through Carolina Biological, which also features free lesson plans and activities and e-mail newsletters. WARD's Natural Science (2007) has served the science education community since 1862, and offers an array of teaching supplies from middle school and high school level to needs for college, homeschool studies, and hobbies. Their animal science materials include bovine, sheep and swine models and organs.

Animal Protection Organizations as a Resource

Many of the animal protection organizations have some involvement in the topic of dissection. While such groups are focused on advocating for animals and lobbying against dissection, they have developed informational or teaching resources relating to dissection and in recent years have provided funding for educational product development. Increasingly, these organizations serve as consultants to assist students or teachers interested in using alternative teaching resources, and this is an expanding role for Animalearn (2007). By setting a specific objective of aligning their work with The Science Bank to meet science education standards, Animalearn's rate of requests from teachers for loans of teaching resources greatly increased.

A division of the Humane Society of the United States, termed the National Association for the Advancement of Humane Education (1985) but now the National Association for Humane and Environmental Education (2007), developed and offers an information packet for teachers. It describes thirteen alternative biology projects that are related to the traditional objectives of dissection and animal study but does not involve animal parts. Most of these activities can be conducted with minimal supplies, presumably at little expense, including such things as palpation of human muscle, dissection and microscopic viewing of flower tissue, and labeling of anatomical structures.

In concluding this chapter one cannot help but notice that teaching aids, especially alternatives for dissection, will become increasingly available and they will be more lifelike, appealing, and comprehensive. Clearly, with at least some nudging from school boards and administrators, the rate at which such useful aids are economically available will accelerate.

CHAPTER 11

Locating Teaching Resources and Research Literature on Alternatives

A major consideration in teaching precollege biology education is that many students may have no further formal education later in their lives related to their own bodies, health, and medical care. This point alone underlines the importance of teaching human biology in seventh grade and high school. Exactly how to balance the tradition in biology education of dissecting animals with the current wide availability of outstanding materials on human biology is an emerging issue that has received little attention. The tradition of dissecting animals and studying their bodies originated at a time when they were used as surrogates for humans, which were much less available (see Chapter 2). Now that a sample of models, plastinated parts, and software images depicting human bodies are readily accessible, we can include them in the array of teaching materials, while also bringing in animal information for special features to increase students' interest and provide some comparative contexts. With regard to the latter topic, some ideas that come to mind are models or illustrations comparing structures related to a dog's olfactory acuity with that of humans (dogs being 100,000–1,000,000 times better than humans); the carriage of several fetuses in the uterus of a cat versus the single fetus of humans; and the hearing system of bats that use echolocation with that of humans.

A starting point to assist teachers who have time to explore the various options is to provide easy access to information about the resources that are available. Fortunately, searching is available that efficiently provides basic information on the resources, dealing both with human biology and physiology, as well as animal biology and physiology. Four different search engines are available dealing specifically with teaching resources; these are:

1. NORINA, Norwegian Reference Centre for Laboratory Animal Science and Alternatives: http://oslovet.veths.no/NORINA.
2. EURCA, European Resource Centre for Alternatives to Using Animals: http://www.eurca.org.
3. AVAR Alternatives in Education Database, Association of Veterinarians for Animal Rights: http://www.avar.org/alted/.
4. InterNICHE, International Network for Humane Education: http://www.interniche.org/alt.html.

We offer some comparisons of these databases here. Additionally, at the UC Center for Animal Alternatives (UCCAA), we have created some specialized searching tools and tutorials that increase efficiency and make it easier to search these four databases (http://www.vetmed.ucdavis.edu/Animal_Alternatives/altsearch.htm). For new users, the UCCAA site offers a detailed tutorial for each database to offer help in using these tools. We also describe the use of AltWeb, a Web site focusing on alternatives, and Google Scholar, a powerful search engine. Table 11.1 concisely lists these and other databases for teaching alternatives.

Techniques of plastination, inspired by the displays of von Hagens' BODY WORLDS (von Hagens and Whalley, 2002) and the advent of the National Library of Medicine's Visible Human software have democratized the availability of high quality visual aids of all aspects of the human body. Anyone with access to the Internet can peruse serial slices of bodies or organs of the Visible Human, male or female, some with explanatory text. While these materials comprise a breadth and depth of information that is readily available, so far, little formal discussion has addressed how these materials might pertain to seventh grade life science and health, and how they might be used. Should they supplant or just supplement existing materials? What should be the proportional balance of attention on human versus animal biology? In laboratory exercises, which types of bodies, tissues, and organs should be offered? Now that human materials are abundant, are animal materials—conventionally fetal pigs, rats, or cats— less relevant and pertinent to education in biology for junior high school students?

The revolution of available Web resources delivers teaching materials throughout the world. How do teachers link in some of these sparkling resources and interface them with their teaching outlines and textbooks? The notion of a stand-alone textbook that provides everything begins to seem outdated. Yet, the infinite array of online materials can be overwhelming and systematic methods for presentation that selectively draw from some of the best resources seem not to exist.

Table 11.1 Databases for Teaching Alternatives

Title and URL	Producer	Content
Alternatives to Dissection packet http://www.nahee.org/ ShoppingCart/ AddToShoppingCart. asp?Prod_ID=NAH012 National Association for Humane and Environmental Education http://www.nahee.org/	The Humane Society of the United States	Humane Society: Youth serves as the youth education affiliate of The Humane Society of the United States (HSUS). They have developed and provide teaching materials, professional development, and other support to teachers and humane educators.
Altweb http://altweb.jhsph.edu/	Johns Hopkins Center for Alternatives to Animal Testing	Altweb, the Alternatives to Animal Testing Web Site, was created to serve as a gateway to alternatives news, information, and resources on the Internet and beyond.
AVAR Alternatives in Education Database http://www.avar.org/ alted/	Association for Veterinarians for Animal Rights	AVAR's Alternatives in Education Database contains thousands of entries of alternatives for many levels of education. To locate a particular alternative, type in a key word or phrase in the blank box. The advanced search option allows searching for an alternative by species, medium, discipline, educational level, or any combination of these categories.
CALF Teaching Media Catalog http://www.calf. vetmed.ucdavis.edu/ calf.html http://169.237.113.35/ FMPro?-db=Products. FP5&-lay=CGI&- token=%5BFMP- currenttoken%5D&- format=search.htm&- view	Computer Assisted Learning Facility, School of Veterinary Medicine, University of California, Davis	Focusing on veterinary education and products developed at the University of California, Davis, School of Veterinary Medicine, the catalog includes online products, videos, software, models, and images.

(*continued*)

Table 11.1 (*continued*)

Title and URL	Producer	Content
ERIC http://www.eric.ed. gov/	U.S. Department of Education's Institute of Education Sciences	Education Resources Information Center (ERIC) is a national information system funded by the U.S. Department of Education's Institute of Education Sciences to provide access to education literature and resources.
EURCA http://www.eurca.org Alternatives database http://www.eurca.org/ resources.asp	European Resource Centre for Alternatives in Higher Education, University of Utrecht, University of Edinburgh	EURCA promotes the use of alternatives to using animals, specifically in higher education. The website and database provide a mechanism for dissemination of information about alternatives. As well as full descriptions of the alternatives, this resource also contains reviews commissioned by EURCA, user comments, other reviews and descriptions, and additional educational materials.
GoogleScholar http://scholar.google. com/	Google	Google Scholar provides a way to broadly search for scholarly literature, across multiple disciplines and sources. Searches selectively cover peer-reviewed papers, theses, books, abstracts and articles, from academic publishers, professional societies, preprint repositories, universities and other scholarly organizations.
ILAR http://dels.nas.edu/ ilar_n/ilarhome/ For educators and students http://dels.nas.edu/ilar_ n/ilarhome/educators. shtml	Institute for Laboratory Animal Research	The Institute for Laboratory Animal Research (ILAR) prepares authoritative reports on subjects of importance to the animal care and use community, as well as develops and makes available scientific and technical information on laboratory animals and other biological research resources.

(*continued*)

Table 11.1 (*continued*)

Title and URL	Producer	Content
ILAR Principles and guidelines for the use of animals in precollege education http://dels.nas.edu/ilar_n/ilarhome/Principles_and_Guidelines.pdf		
InterNICHE http://www.interniche.org/index1.html Databases, resources http://www.interniche.org/alt_info.html	InterNICHE	InterNICHE organizes and provides access to publications and searchable databases that provide detailed information on applications, specifications and distributors for several thousand alternative products in all fields of biological science, veterinary and human medicine.
NORINA http://oslovet.veths.no/norina	Laboratory Animal Unit, Norwegian School of Veterinary Science, Oslo, Norway	NORINA is an English-language database containing detailed information on about 4,000 audiovisual aids and other alternatives for human and animal biology that may be used as alternatives or supplements to the use of animals in teaching and training.

Despite the availability of some extraordinary free educational materials for biology, most of the higher-quality Web-based resources still operate in a proprietary/commercial context and need to be purchased, as do textbooks. With the flood of Web materials, it seems likely that the proprietary products will be placed in a less central role and we will soon see a more diversified approach with availability of learning materials without cost to the user. Already, the availability of some free Web-based content and virtual dissections, plus Internet-accessible free loan programs are bringing free resources more to the forefront. This is the system that exists on most college campuses with regard to scientific journals where students, through their own computers, have immediate access to full text research papers and are able to print most of them without ever holding the journals, or walking through the library book stacks. The access to journals requires an institutional subscription, campus connection, and password. It would seem as though a similar arrangement could be made at modest cost for precollege student access to Web-based learning aids.

In the 1960s, extensive science resource centers in each county served teachers, delivering to them the laboratory materials they needed each week. This process was guided by standard curricular materials and formal lesson plans that made specific suggestions for each lesson. The supportive structure of resource centers with the assistance in preparing teaching laboratories each week has now been dismantled. However, in their place we have a growing supply of outstanding materials. Many of them are freely available, but they are not packaged in an orderly handbook that offers a teacher a suggested schedule and lesson plan with activities from which to select. Rather, the teacher must wade through everything, and then figure out what is worthwhile and will fit within the time and classroom context.

A paradox discussed in Chapter 6 is that animal materials can be used in precollege classrooms without any institutional review or oversight, even though the same procedures if conducted in a college or university setting would require institutional approval.

DATABASES AND SEARCH ENGINES ON TEACHING RESOURCES

Four databases have been established specifically to make it easier for teachers to locate the types of resources that they are seeking. These databases emphasize proprietary products that are available for purchase. In addition, the broad search engine, Google, can be used to conduct searches for these resources. Except for NORINA, Web-based software appears to be underrepresented in the four alternatives databases. For this reason, we recommend a strategy that uses both databases and search engines.

NORINA (http://oslovet.veths.no/NORINA)

The most comprehensive database on teaching resources, NORINA has about 4,000 entries. An individual record on each item provides all the basic information that a prospective user would hope to find, including a full description, ordering information, price, and in many cases, a picture of the product. Proceeding to the NORINA Web site where the resources are cataloged, we find resources for teaching in nearly thirty categories including anatomy, dissection, histology, and organs. Also covered are anesthesia, pathology, pharmacology, physiology, and surgery. Most relevant for precollege are the categories of anatomy (or dissection), preserved materials, and physiology. Species other than humans that are included among the anatomy resources are bird, cat, chick, clam, crayfish, dog, earthworm, fetal pig, fish, grasshopper, horse, mammals, perch, pigeon, rabbit, rat, sheep, and starfish. Organs such as brain, ear, eye, heart, kidney, and liver are presented. Among the systems, cardiovascular, digestive, muscular, reproductive, and respiratory systems are featured. Materials are categorized in nearly thirty types including computer programs, CD-ROMs, films, slides, charts/diagrams, display mounts, review sheets, skeletons, and videotapes.

To provide teachers even more efficient access to the NORINA entries, a searching Web site with a tutorial offers point-and-click menu grids oriented to particular topics that teachers may need under the categories of anatomy, dissection, histology, organs, and physiology, as well as the medically-related topics (UC Center for Animal Alternatives: UCCAA, 2007). Subcategories available for scrolling in the pop-up menus include breakdowns by species or type of tool, such as CD-ROM, preserved specimens, or teaching model. The UCCAA Web site grids provide the user with a manageable subset of entries in a specified category. When an item is selected in this database, it links to the standardized detailed record described above, specifying the program name, category, type, description, source, telephone, fax, price, computer type, year of version, author, comments and references, e-mail, and Web site. The record lists important information about where the item has been listed under a different name, where it can be accessed at no charge, or borrowed in a loan program. It provides advice regarding hardware requirements for running some of the software.

In a sample search in NORINA for "human anatomy Web pages," nine resources were identified.

EURCA (http://www.eurca.org/)

A much smaller and more selective database is offered by EURCA, which includes categories of dissection, anatomy, physiology, and features models, computer programs, and videos. Among the anatomy resources, only one highlights humans. However, several of the physiology laboratories include experiments with humans. EURCA provides a very full description of the product with much more details than found in the other databases, with the exception of NORINA. In addition to the in-depth information provided, an advantage of the EURCA site is that roughly half of the items listed are reviewed by users, so the reader can get a feeling for the particular advantages and disadvantages of the product. Searching EURCA for "dissection" yielded ten resources. Searching EURCA for computer program (CD-ROM, floppy disk) yielded fifty resources.

The UCCAA site for EURCA provides a point-and-click grid, with direct links to records on the following topics: dissection, anatomy, physiology, model, computer program, and video. When an item is selected, the link leads to a brief listing on EURCA, providing the title, a summary description, the supplier and price, a summary of reviews, and links to full texts of reviews. A link offers a full description of the resource type, educational level, language, detailed description, target group, the animal experiment replaced, the supplied support materials, notes on teaching use, practical time, analysis time, author, year of version, NORINA reference number (linked), InterNICHE loan (linked), listing in alternatives databases, supplier (linked), and full ordering information (linked). Users are offered an opportunity to provide comments about these alternatives.

AVAR Alternatives in Education Database (http://www.avar.org/alted/)

A long list of topics can be searched on the database of teaching alternatives offered by AVAR that appears to comprise around 6,000 records. The default simple search allows selection by the most restrictive "match exact phrase," the broadest "match any word," and the most similar to Google, "match all words." There is also an advanced search option, allowing the search to be narrowed by discipline (anesthesia, brain, dissection, intubation, surgery, ultrasound), medium (slides, CD-ROM, DVD, model, simulator, software), and education level (primary, high school, undergraduate, graduate, and professional). Searching the Alternatives in Education Database for "cat dissection" as an exact phrase retrieves twelve; searching for either word retrieves over 500, as does searching for both words in any order. The advanced search, however, allows the search results to be greatly narrowed and more relevantly focused, by limiting the results to "software" and "primary" school-age, for example. The UCCAA site suggests appropriate terms for searching that can be combined in searching the AVAR database, including animal species, organs/parts, disciplinary categories, and resource types. Selecting a linked item from this AVAR Alternatives in Education Database generates a standard record, including the title, medium, discipline, species, educational level, cost, source (linked), catalog, and notes.

InterNICHE (http://www.interniche.org/)

The fourth database is maintained by InterNICHE. It focuses on providing lists of alternatives for teaching in the biological sciences and medical and veterinary education, and offers categories of anatomy, clinical skills and surgery, physiology, and miscellaneous topics (Jukes and Chiuia, 2003). The UCCAA grid for this site gives point-and-click access to the categories of anatomy, clinical skills and surgery, physiology, and miscellaneous. InterNICHE organizes terms within each category under subtopics, including CD-ROMs, models and simulators, videos, and books. When an item in this database is selected, it links directly to the Web site of the provider of the product. When EURCA reviews are available, direct links are provided to the EURCA reviews.

Searching InterNICHE under "anatomy" yielded twenty records for CD-ROMs, two for models, and eleven for videos. In addition to identifying alternatives products, InterNICHE organizes a Loan Library, as well as listing other alternatives loans systems. Borrowing from these loan libraries allows a teacher to actually use a model or software before purchasing.

AltWeb (http://altweb.jhsph.edu/)

The Johns Hopkins University Center for Alternatives in Animal Testing (CAAT) offers a specialized Web site and search engine that focuses specifically on alternatives. AltWeb is described as a global clearinghouse for information on alternatives to animal testing. While most of the resources focus on animal use in research and testing, some deal with teaching resources. Also included on this site is a tutorial to assist in searching. Key words that are useful for accessing information for teaching are "education resources" and "dissection."

Google and Google Scholar (http://scholar.google.com)

With Google, an ultimate search engine, one can easily be overwhelmed by thousands of hits when searching these topics. For example, a search in Google for "cat dissection software" yielded 334,000 hits, but the initial pages included valuable resources, and there was no need to proceed further. Nonetheless, the initial few pages of a search may provide useful information on resources that otherwise would not easily come to light. Web-based resources are quickly accessed in Google, including pages with selected links to outstanding resources.

Working within Google Scholar provides access to a more credible pool of information, drawing from published scientific literature that is available in the public domain. Sources located on Google Scholar are more likely to be authoritative and research-based, identified as a result of a review or study. Using the same keywords, "cat dissection software," in Google Scholar will retrieve very different results from those found in Google; most obvious is that the links will be to papers that have been published on specific cat dissection software as opposed to links to the products themselves.

TEACHING RESOURCES ON HUMANS

Extensive and numerous resources are available that pertain to humans; not surprisingly, most of the resources available relate to humans, not animals. On August 2, 2006, the NORINA database for either anatomy or dissection on humans yielded 1,313 records, as compared with 143 records for cats, 56 records for fetal pigs, and 212 records for frogs. Some of these resources on humans are highly technical and designed for medical students, physicians, or medical specialists, but others would be appropriate for use by precollege students. Google can also be used to access many resources on humans that can be reviewed to identify ones that may meet the needs of a particular teacher.

Using Google, one can quickly access specialized lists of links to teaching resources. The individual sites can then be perused to identify sites of particular usefulness. Some examples follow:

http://www.folkstone.ca/anatomist/anatomy/teaching.html;

http://www.anatomyatlases.org/;

http://www.csun.edu/science/biology/anatomy/anatomy.html;

http://www.mic.ki.se/Anatomy.html;

http://schoolscience.rice.edu/duker/sciteanatomyres.html;

http://www.kidinfo.com/Health/Human_Body.html;

http://edtech.kennesaw.edu/web/humanbo.html.

What is noticeable about these sites is that they typically feature very little information concerning animals, even though the tradition in precollege classrooms includes an animal emphasis.

Some of the individual sites that seemed prospective are these:

http://www.accessexcellence.org/RC/VL/xrays/index.html;

http://www.medtropolis.com/VBody.asp;

http://innerbody.com/index.html (product by Google);

http://www.bbc.co.uk/science/humanbody/ (product by BBC);

http://www.fi.edu/biosci/systems/systems.html (product by Franklin Institute).

TEACHING RESOURCES ON ANIMALS

Despite the numbers of resources that pertain directly to humans, the four databases of teaching resources conventionally are viewed as paths to animal alternatives. Almost 4,000 resources are available on NORINA, and a large number also on AVAR. Since one can make the argument that ideal resources are not available for precollege instruction in biology and health, it seems almost paradoxical that such a huge number of teaching resources exists. For example, two Web sites for frog dissection are http://frog.edschool.virginia.edu/ and http://froggy.lbl.gov/virtual/ (product by Lawrence Berkeley National Laboratory).

RESEARCH LITERATURE ON ALTERNATIVES FOR TEACHING AND DISSECTION

Having perused the resources available for teaching biology to precollege students, and in particular seventh graders, it may seem hard to understand why alternative teaching resources have not replaced dissection. Perhaps

Table 11.2 Research Bibliography on Dissection and Alternatives in Teaching

History of Dissection and Science Education: Books and Articles

Carlino, A. (1999). *Books of the body: Anatomical ritual and Renaissance learning* (J. Tedeschi and A.C. Tedeschi, translator). Chicago: University of Chicago Press.

DeBoer, G.E. (1991). *A history of ideas in science education: Implications for practice.* New York: Columbia University.

DeBoer, G.E. (2000). Scientific literacy: Another look at its historical and contemporary meanings and its relationship to science education reform. *Journal of Research in Science Teaching* 37:582–601.

French, R. (1999). *Dissection and vivisection in the European Renaissance.* Aldershot, England: Ashgate.

French, R. (2000). *Ancients and moderns in the medical sciences: From Hippocrates to Harvey.* Aldershot, England: Ashgate.

Huxley, T.H. (1854/1902). *Science and education.* New York: P. F. Collier & Son.

Klestinec, C. (2004). A history of anatomy theaters in sixteenth century Padua. *Journal of the History of Medicine* 59(3): 375:412.

Library of Congress Research Centers. (1990). The crisis in science education. Tracer Bullet 90–95. Washington, DC: The Library of Congress. Web site accessed May 23, 2007, http://www.loc.gov/rr/scitech/tracer-bullets/crisis-edutb.html

Maehle, A.-H. (1993). The ethical discourse on animal experimentation, 1650–1900. *Clio Medica* 24:203-251.

Marshall, T. (1995). *Murdering to dissect: Grave-robbing, Frankenstein, and the anatomy literature.* Manchester, England: Manchester University Press.

O'Malley, C.D. (1964). *Andreas Vesalius of Brussels: 1514–1564.* Berkeley: University of California Press.

Richardson, R. (2000). *Death, dissection and the destitute.* Chicago: University of Chicago Press.

Richardson, R. (2006). Human dissection and organ donation. *Mortality* 11:151–165.

Rudolph, J.L. (2002). *Scientists in the classroom: The cold war reconstruction of American science.* New York City: Palgrave.

Sappol, M. (2002). *A traffic of dead bodies: Anatomy and embodied social identity in nineteenth-century America.* Princeton: Princeton University Press.

Sawday, J. (1995). The body emblazoned: Dissection and the human body in Renaissance culture. London: Routledge.

Scientific Journals and Conference Proceedings on Alternatives

ALTEX (Alternativen zu Tierexperimenten)

ATLA (Alternatives to Laboratory Animals)

Animal Welfare

(continued)

Table 11.2 (*continued*)

JAAWS (Journal of Applied Animal Welfare Science)

World Congress on Alternatives and Animal Use in the Life Sciences, proceedings

Books on Dissection and Alternatives

Balcombe, J. (1996). *Attitudes toward dissection: annotated list of studies.* Washington, DC: Humane Society of the United States.

Balcombe, J. (2000). *The use of animals in higher education: Problems, alternatives, and recommendations.* Washington, DC: Humane Society Press. http://www.hsus.org/web-files/PDF/ARI/ARIS_The_Use_Of_Animals_In_Higher_Education.pdf

Francione, G.L. and Charlton, A.E. (1992). *Vivisection and dissection in the classroom: A guide to conscientious objection.* Jenkintown, PA: American Anti-Vivisection Society.

Jukes, N. and Chiuia, M. (2003). *From guinea pig to computer mouse: Alternative methods for a progressive, humane education* (2nd ed.). Leicester, England: InterNICHE.

Orlans, F.B. (1993) *The use of animals in education.* In: *In the name of science: Issues in responsible animal experimentation.* Oxford: Oxford University Press.

Pedersen, H. (2002). *Humane education: Animals and alternatives in laboratory classes. Aspects, attitudes and implications.* Stockholm: Stifelsen Forskning utan djurförsök. http://www.interniche.org/thesis_helena.pdf

Simon, L. (1995). *Beyond dissection: Innovative teaching tools for biology education.* Boston, MA: Ethical Science Education Coalition.

Position Papers

Balcombe, J. (1999). Comparative studies of dissection and other animal uses in education. Washington, DC: Humane Society of the United States. http://www.hsus.org/animals_in_research/animals_in_education/comparative_studies_of_dissection_and_other_animal_uses.html

InterNICHE Position papers and Policies, *located in* From guinea pig to computer mouse http://www.interniche.org/news.html#book

NORINA Guidelines for the care and use of animals in research. http://oslovet.veths.no/guidelines

Responsible Use of Live Animals and Dissection in the Science Classroom. NSTA Position Statement. National Science Teachers Association (NJ1), 2005. http://www.nsta.org

Images

Dream Anatomy. History of Medicine, National Library of Medicine. http://www.nlm.nih.gov/exhibition/dreamanatomy/index.html

(*continued*)

Table 11.2 (*continued*)

Images from the history of medicine. History of Medicine, National Library of Medicine. http://wwwihm.nlm.nih.gov/

Prints and Photographs Online Catalog. Prints and Photographs Division, Library of Congress. http://lcweb2.loc.gov/pp/pphome.html

Recent Articles on Dissection and Science Education: Professional Schools

Dyer, G.S.M. and Thorndike, M.E.L. (2000). Quidne mortui vivos docent? The evolving purpose of human dissection in medical education. *Academic Medicine* 75(10):969–979.

Hart, L.A., Wood, M.W., and Weng, H-Y. Mainstreaming Alternatives in Veterinary Medical Education: Resource Development and Curricular Reform. *Journal of Veterinary Medical Education* 2005 32(4):473–480.

Hedlund, C.S., Hosgood, G., and Naugler, S. (2002). Surgical education: attitudes toward animal use in teaching surgery at Louisiana State University. *Journal of Veterinary Medical Education* 29(1):50–55.

Knezevic G. (2004). Animal use in higher education in the SEE (South Eastern Europe) region. *ALTEX* 21(2):102–104.

Nobis N.J (2002). Animal dissection and evidence-based life-science and health-professions education. *Journal of Applied Animal Welfare Science* 5(2):157–161.

Patronek, G.J. and Rauch, A.J. (2007). Systematic review of comparative studies examining alternatives to the harmful use of animals in biomedical education.

Journal of the American Veterinary Medical Association 230(1):37–43.

Provo, J., Lamar, C. and Newby, T. (2002). Using a cross-section to train veterinary students to visualize anatomical structures in three dimensions. *Journal of Research in Science Teaching* 39(1):10–34.

Smith, A. and Allen, T. (2005). The use of databases, information centres and guidelines when planning research that may involve animals. *Animal Welfare* 14(4):347–359.

Smith A. and Smith K. (2004). Guidelines for humane education: Alternatives to the use of animals in teaching and training. ATLA, 32 Supplement 1:29–39. http://www.worldcongress.net/2002/proceedings/SP4 Smith.pdf

Recent Articles on Dissection and Science Education: Precollege

Allchin, D. (2005). "Hands-Off" Dissection? *American Biology Teacher* 67(6):369.

Fabian, C.A. (2004). Evolutionary biology digital dissection project: Web-based laboratory learning opportunities for students. *American Biology Teacher* 66(2):128.

(*continued*)

Table 11.2 (*continued*)

Franklin, S., Peat, M., and Lewis, A. (2002). Traditional versus computer-based dissections in enhancing learning in a tertiary setting: A student perspective. *Journal of Biological Education* 36(3):124–129.

Madrazo, G.M. (2002). The debate over dissection: Dissecting a classroom dilemma. *Science Educator* 11(1):41–45.

Orlans, F.B. (1995). Investigator competency and animal experiments: Guidelines for elementary and secondary education. *Lab Animal* 24(9):29, 32–34.

Roy, K. (2004). Responsible Use of Live Animals in the Classroom. *Science Scope* 27(9):10–11.

what is lacking is a clear match between the lesson plans and the materials to be used in laboratories. The burden is on the teacher to find the specific resources that link up with the teaching objectives and the constraints imposed by the classroom layout, computer availability, and time available. Recognizing this challenge that teachers face, Animalearn (2007) with its Science Bank, has been increasingly serving as a consultant to teachers with needs for teaching resources. Teachers with specific teaching objectives are assisted in identifying appropriate products that fit the lesson plan.

Much of the research literature relating to dissection has been referenced within the chapters of this book. As a low-profile topic in recent years, it can be challenging to quickly identify key resources. With that in mind, we have listed some of the most relevant references in Table 11.2. In addition, a new Web site has been developed as a companion to this book: http://www.vetmed.ucdavis.edu/Animal_Alternatives/appendices.html.

While the eight tables in the book and on the Web site provide links to online resources and information relevant to the topics throughout this book (testing, standards and frameworks, laws and regulations, organizations, loan libraries, teaching resources, databases, and bibliography), the Web site also includes a table of search templates. These search templates are stored search strategies embedded in the highlighted links that will allow the interested user to conduct point-and-click searches in Google Scholar, the general search engine described above, and in ERIC (Education Resources Information Center, 2007), an educational database. The templates direct searches in Google Scholar and ERIC, on the subjects of testing, standards, legislation, organizations, loan libraries, and teaching resources. For example, by clicking on the "NCLB and testing" link, a search will automatically be run in the ERIC database and the most recent publications on that topic will be identified and retrieved. The number of citations yielded with some of these general searches can be very large, but users can add additional terms to narrow the search. An additional useful Web-based tool is the Web site, Educational Resources for Teaching, of the UC Center for Animal Alternatives: http://www.vetmed.ucdavis.edu/Animal_Alternatives/dissection.htm. This site includes a variety of additional search grids related to teaching resources.

References

Achieve. (2007). Closing the expectations gap: 2007, an annual 50-state progress report. Web site accessed on June 30, 2007, http://www.achieve.org/files/50-state-07-Final.pdf

ACT. (2006). ACT national score report press release: Aug 16. Web site accessed on July 5, 2007, http://www.act.org/news/releases/2006/ndr.html

A.D.A.M. (2007a). A.D.A.M. essentials high school suite, starting at 164.95. Web site accessed on February 10, 2007, http://www.adam.com/Our_Products/School_and_Instruction/Educators/High_School/essentials.html

———. (2007b). A.D.A.M. our products. Web site accessed on July 6, 2007, http://www.adam.com/Our_Products/School_and_Instruction/index.html

Adkins, J., and Lock, R. (1994). Using animals in secondary education—a pilot survey. *Journal of Biological Education* 8:48–52.

Ahlgren, A. (1996). How standards fit within the framework of science education reform. In: *Issues in science education* (J. Rhoton and P. Bowers, Eds.), pp. 40–45. Arlington, VA: National Science Teachers Association.

Akpan, J.P., and Andre, T. (2000). Using a computer simulation before dissection to help students learn anatomy. *Journal of Computers in Mathematics and Science Teaching* 19:297–313.

American Alliance for Health, Physical Education, Recreation and Dance. (2007). Health education standards. Web site accessed on July 5, 2007, http://www.aahperd.org/aahe/pdf_files/standards.pdf

American Anti-Vivisection Society (AAVS). (2007). The science bank: Education for the future. Web site accessed on July 5, 2007, http://www.aavs.org/images/sciencebank.pdf

American Association for the Advancement of Science. (1989). *Science for all Americans.* New York: Oxford University Press.

———. (1993). *Benchmarks for science literacy.* New York: Oxford University Press.

———. (2005). Science literacy for a changing future: Mathematics, natural sciences, social sciences, technology. *2061 Today* 15(1):1–8. Web site accessed on July 5, 2007, http://www.project2061.org/publications/newsletter/default.htm

American Association for Laboratory Animal Science. (2005). Position statement on humane care and use of laboratory animals: Use of animals in pre-college education. Web site accessed on October 25, 2005, http://www.aalas.org/association/position_statements.asp

American Physiological Society (1994). Carolina Biological cleared on animal cruelty charges. *The Physiologist* 37:200. Web site accessed on July 8, 2007, http://www.the-aps.org/publications/tphys/legacy/1994/issue4/197.pdf

———. (2005). APS position statement and rationale on the use of animals in teaching. *The Physiologist* 48:206–208. Web site accessed on July 10, 2007, http://www.the-aps.org/pa/resources/policyStmnts/paPolicyStmnts_teaching.htm

———. (2007). The value of animals in teaching. Web site accessed on July 8, 2007, http://www.the-aps.org/pa/policy/animals/AnimalLabValue.htm

Animalearn. (2007). The science bank: Alternatives to animal use in education. Web site accessed on July 5, 2007, http://animalearn.org/home.php

Animal Place. (2007). Animal Place. Web site accessed on July 6, 2007, http://www.animalplace.org

Arizona Department of Education. (1997). Comprehensive health education/physical activity standards. Web site accessed on June 22, 2007, http://www.ade.state.az.us/standards/health/

Arluke, A. (1996). From apprehension to fascination with "Dog Lab": The use of absolutions by medical students. *Journal of Contemporary Ethnography* 25:201–225.

———. (2004). The use of dogs in medical and veterinary training: Understanding and approaching student uneasiness. *Journal of Applied Animal Welfare Science* 7:197–204.

Ascione, F.R. (1997). Humane education research: Evaluating efforts to encourage children's kindness and caring toward animals. *Genetic, Social and General Psychology Monographs* 123:57–77.

Ascione, F.R., and Arkow, P. (1999). *Child abuse, domestic violence, and animal abuse: Linking the circles of compassion for prevention.* West Lafayette, IN: Purdue University Press.

Ascione, F.R., and Weber, C.V. (1996). Children's attitudes about the humane treatment of animals and empathy: One-year follow up of a school-based intervention. *Anthrozoos* 9:188–195.

Asimov, N. (2007a). "Huge" drop in graduations. *San Francisco Chronicle* May 9:B3.

———. (2007b). Justices back special ed parents. *San Francisco Chronicle* May 22:B1, B8.

Asimov, N., and Wallack, T. (2007a). The teachers who cheat. *San Francisco Chronicle* May 13:A1, A10.

———. (2007b). Stakes too high to just check erasures, experts say. *San Francisco Chronicle* May 13:A10.

Asma, S.T. (2001). *Stuffed animals & pickled heads: The culture and evolution of natural history museums.* Oxford: Oxford University Press.

Association for Assessment and Accreditation of Laboratory Animal Care International. (2007). AAALAC International. Web site accessed on August 12, 2007, http://www.aaalac.org/

Association of Veterinarians for Animal Rights (AVAR). (2007). AVAR. Web site accessed on July 3, 2007, http://www.avar.org/

Ausubel, D.P. (1968). *Educational psychology: A cognitive view.* New York: Holt, Rinehart and Winston.

Baker, D.W., Wolf, M.S., Feinglass, J., Thompson, J.A., Gazmararian, J.A., and Huang, J. (2007). Health literacy and mortality among elderly persons. *Archives of Internal Medicine* 167:1503–1509.

Balcita, A.M., Carver, D.L., and Soffa, M.L. (2002). Shortchanging the future of information technology: The untapped resource. *SIBCSE Bulletin* 34:32–35.

Balcombe, J. (1996, Summer). Dissection and the law: Dissecting dissection. *AV Magazine* 18–21.

———. (1997). Student/teacher conflict regarding animal dissection. *The American Biology Teacher* 59:22–25.

———. (2000). *The use of animals in higher education: Problems, alternatives, and recommendations.* Washington, DC: Humane Society Press.

———. (2001). Dissection: The scientific case for alternatives. *Journal of Applied Animal Welfare Science* 4:117–126.

———. (2002, Fall). High school dissection: Evolution but not yet revolution. *AV Magazine* 19–21.

Balcombe, J.P. (2004). Medical training using simulation: Toward fewer animals and safer patients. *ATLA Supplement 1* 32:553–560.

Bambach, C. (2002). Anatomy in the Renaissance. In: *Timeline of Art History.* New York: The Metropolitan Museum of Art, 2000-. Website accessed August 9, 2007, http://www.metmuseum.org/toah/hd/anat/hd_anat.htm

Barr, G., and Herzog, H. (2000). Fetal pig: The high school dissection experience. *Society & Animals* 8:53–70.

Barrett, C.R. (2005, February 24). Educational complacency will make U.S. feel the pain. *USA Today* 13A.

Bell, B.F. (1981). When is an animal, not an animal? *Journal of Biological Education* 15:213–218.

Belt, E. (1955). *Leonardo the anatomist.* Lawrence: University of Kansas Press.

Biological Sciences Curriculum Study. (1963). *Biological science: An inquiry into life* (BSCS Yellow Version). New York: Harcourt Brace; *Biological science: Molecules to man* (BSCS Blue Version). Boston: Houghton Mifflin; *High school biology* (BSCS Green Version). Chicago: Rand McNally.

Biskupic, J. (2007, May 22). Parents of disabled kids can represent selves in court. *USA Today* 2A.

Bjerke, T., Odegardstuen, T.S., and Kaltenborn, B.P. (1998). Attitudes toward animals among Norwegian adolescents. *Anthrozoos* 11:79–86.

BODY WORLDS. (2002). Gunther von Hagens' BODY WORLDS: The anatomical exhibition of real human bodies. Program brochure. London.

———. (2007a). Gunther von Hagens' BODY WORLDS: The anatomical exhibition of real human bodies. Web site accessed on July 6, 2007, http://www.bodyworlds.com/en.html

———. (2007b). Gunther von Hagens—A life in science. Web site accessed on July 8, 2007, http://www.bodyworlds.com/en/gunther_von_hagens/life.html

Bonner, J.J. (2004). Changing strategies in science education. *Science* 306:228.

Bowd, A.D. (1993). Dissection as an instructional technique in secondary science: Choice and alternatives. *Society & Animals* 1:83–88.

Bryant, B. (1985). The neighborhood walk: Sources of support in middle childhood. *Monographs of the Society for Research in Child Development* 50(3, Serial No. 210): 34–44.

Bybee, R.W. (Ed.) (2002). *Learning science and the science of learning.* Arlington, VA: NSTA Press.

California Aggie. (2007, May 23). Judge dismisses lawsuit seeking standardized testing in Spanish, p. 8.

California Science Teachers Association (CSTA). (2007). CSTA. Web site accessed on July 5, 2007, http://www.cascience.org/scienceeducationsuppliers.html

California State Board of Education. (2007a). Biology/life sciences—Grades nine through twelve. Web site accessed on July 4, 2007, http://www.cde.ca.gov/be/st/ss/scbiology.asp

———. (2007b). Science content standards—Grade seven. Web site accessed on July 4, 2007, http://www.cde.ca.gov/be/st/ss/scgrade7.asp

California State Board of Education: Curriculum Development and Supplemental Materials Commission. (2003). *Health framework for California public schools.* Sacramento, CA: California Department of Education. Web site accessed on July 5, 2007, http://www.cde.ca.gov/re/pn/fd/health-frame-pdf.asp

Campbell, J.A., and Meyer, S.C. (Eds.) (2004). *Darwinism, design, and public education.* East Lansing, MI: Michigan State University.

Carey, S. (1985). *Conceptual change in childhood.* Cambridge, MA: MIT Press.

Carlino, A. (1999). *Books of the body: Anatomical ritual and Renaissance learning* (J. Tedeschi and A.C. Tedeschi, Trans.). Chicago: University of Chicago Press.

Carolina Biological Supply Company. (2004a). Dissection fact sheet. Web site accessed on August 28, 2006, http://www.carolina.com/general/departments/dissection.asp

———. (2004b). Science & technology concepts for children: STC/MS. Web site accessed on February 10, 2007, http://www.carolina.com/stcms/overview.asp

———. (2004c). Living materials care and handling guide. Web site accessed on February 10, 2007, http://www.carolina.com/STC/PDF/Livingmatcareguide.pdf

———. (2007). Carolina. Web site accessed on July 10, 2007, http://www.carolina.com/

Chapman, E. (2003). Alternative approaches to assessing student engagement rates. *Practical Assessment, Research & Evaluation* 8:13. Web site accessed on February 2, 2005, http://pareonline.net/getvn.asp? v=138&n=13

Choosing the Best. (2007). The leader in abstinence education. Web site accessed on June 15, 2007, http://www.choosingthebest.org/

Conant, J.B. (1947). *On understanding science.* New Haven, CT: Yale University Press.

Conati, C., and Klawe, M. (2002). Socially intelligent agents in educational games. In: *Socially intelligent agents—Creating relationships with computers and robots* (K. Dautenhahn, A. Bond, D. Canamero, and B. Edmonds, Eds.), 9 pp. Boston: Kluwer Academic Publishers.

Conati, C., and Zhou, X. (2002). Modeling students' emotions from cognitive appraisal in educational games. In: *ITS 2002* (S.A. Cerri, G. Bouarderes, and F. Paraguacu, Eds.), LNCS 2363:944–954. Berlin: Springer-Verlag.

Craven, J. (1999). Puppy love. *Times Educational Supplement* August 13:14–15.

Cunningham, A. (1997). *The anatomical Renaissance: The resurrection of the anatomical projects of the ancients.* Aldershot, England: Scolar Press.

Daly, B., and Morton, L.L. (2006). An investigation of human–animal interactions and empathy as related to pet preference, ownership, attachment, and attitudes in children. *Anthrozoos* 19:113–127.

Davis, L.A., and Gibbin, R.D. (Eds.) (2002). *Raising public awareness of engineering.* The National Academies Press. Web site accessed on August 12, 2007, http://books.nap.edu/catalog.php?record_id=10573

Darwin, C. (1859/2003). *The origin of species by means of natural selection.* New York: Alfred A. Knopf.

DeAguilera, M., and Mendiz, A. (2003). Video games and education (Education in the face of a "Parallel School"). *ACM Computers in Entertainment* 1(1): 14 pp.

DeBoer, G.E. (1991). *A history of ideas in science education: Implications for practice.* New York: Columbia University.

———. (2000). Scientific literacy: Another look at its historical and contemporary meanings and its relationship to science education reform. *Journal of Research in Science Teaching* 37:582–601.

Delacote, G. (1998). Putting science in the hands of the public. *Science* 280:2054–2055.

Descartes, R. (1637). *Discourse on the method of rightly conducting the reason.* Descartes: A new approach. Web site accessed on August 12, 2007, http://www.philosophypages.com/hy/4b.htm#anim

Dewey, John. (1916a). *Democracy and education: An introduction to the philosophy of education.* New York: Macmillan.

———. (1916b). *Experience and education.* New York: Collier.

Dickson, C. (2001). Early modern anatomies. In: *Experience and Experiment in Early Modern Europe: A Summer 2001 NEH Institute.* Web site accessed on December 4, 2007, http://www.folger.edu/html/folger_institute/experience/mechanical_dickson.htm

Douglass, C., and Glover, R. (2003). Plastination: Preservation technology enhances biology teaching. *American Biology Teacher* 65:503–510.

Driver, R., and Bell, B. (1986). Students' thinking and the learning of science: A constructivist view. *School Science Review* 67:443–456.

Driver, R., Squires, A., Rushworth, P., and Wood-Robinson, V. (1994). *Making sense of secondary science: Research into children's ideas.* New York: London.

Duschl, R.A., Schweingruber, H.A., and Shouse, A.W. (Eds.) (2007). *Taking science to school: Learning and teaching science in grades K-8.* National Academies Press. Web site accessed on May 22, 2007, http://www.nap.edu/catalog/11628.html

Ebenezer, J.V., and Zoller, U. (1993). Grade 10 students' perceptions of and attitudes toward science teaching and school science. *Journal of Research in Science Teaching* 30:175–186.

Education Resources Information Center. (2007). ERIC. Web site accessed on August 12, 2007, http://www.eric.ed.gov/

Einsteins-Emporium. (2006). Animal specimens. Web site accessed on February 10, 2007, http://www.einsteinsemporium.com/life/specimens/ls000.htm

European Resource Centre for Alternatives to Using Animals (EURCA). (2007). EURCA. Web site accessed on July 10, 2007, http://www.eurca.org/

European Union. (2007). *Fourth report on the statistics on the number of animals used for experimental and other scientific purposes in the member states of the European Union.* Web site accessed on July 10, 2007, http://www.animalexperimentfacts.info/assets/studies/Animal%20use%20EC%202005.pdf

Ezarik, M. (2006). The textbook adoption mess—and what reformers are doing to fix it. Web site accessed on March 20, 2007, http://districtadministration.ccsct.com/pageprint.cfm?p=1022

Fawver, A.L., Branch, C.E., Trentham, L., Robertson, B.T., and Beckett, S.D. (1990). A comparison of interactive videodisc instruction with live animal laboratories. *American Journal of Physiology* 259(*Advances in Physiology Education* 4):S11–S14.

Feller, B. (2006). Study says nations whose kids like math the most score worst. *San Francisco Chronicle* October 18:A6.

Fessler, D.M.T., Pillsworth, E.G., and Flamson, T.J. (2004). Angry men and disgusted women: An evolutionary approach to the influence of emotions on risk taking. *Organizational Behavior and Human Decision Processes* 95:107–123.

Forrest, B., and Gross, P.R. (2004). *Creationism's Trojan horse: The wedge of intelligent design.* New York: Oxford University Press.

Fowler, H.S., and Brosius, E.J. (1968). A research study on the values gained from dissection of animals in secondary school biology. *Science Education* 52:55–57.

Fox, C.H., and Benton, C. (1987). Formaldehyde: The fixative. *Journal of Histotechnology* 10:199–201.

French, R. (1999). *Dissection and vivisection in the European Renaissance.* Aldershot, England: Ashgate.

———. (2000). *Ancients and moderns in the medical sciences: From Hippocrates to Harvey.* Aldershot, England: Ashgate.

Furnham, A., and Heyes, C. (1993). Psychology students' beliefs about animals and animal experimentation. *Personality and Individual Differences* 1:1–10.

Furnham, A., and Pinder, A. (1990). Young people's attitudes to experimentation on animals. *The Psychologist* 10:444–448.

Gathright, A. (2003, October 5). Engrossing education in San Jose: Kids learn bodily functions with glee. *San Francisco Chronicle* A1, 24.

Gelles, K. (2006, July 12). Gender shifts. *USA Today* 11A.

Glasgow, N.A. (1996). *Doing science: Innovative curriculum for the life sciences.* Thousand Oaks, CA: Corwin Press.

Greene, J.P. (2002). High school graduation rates in the United States. Manhattan Institute for Policy Research. Web site accessed on February 2, 2007, http://www.manhattan-institute.org/html/cr_baeo.htm

Hagelin, J., Hau, J., and Carlsson, H.-E. (1999). Undergraduate university students' views of the use of animals in biomedical research. *Academic Medicine* 74:1135–1137.

Hamm, T., and Blum, J. (1992). The humane use of animals in teaching. *American Association for Laboratory Animal Science* 31:20–25.

Hancocks, D. (2001). *A different nature: The paradoxical world of zoos and their uncertain future.* Berkeley: University of California Press.

Hart, L.A. (1995). Animal alternatives in precollege education: When? *In Vitro Toxicology* 8:213–214.

Hart, L.A., and Wood, M.W. (2004). Uses of animals and alternatives in college and veterinary education at the University of California, Davis: Institutional commitment for mainstreaming alternatives. In: *Alternatives and Animal Use in the Life Sciences* (A. Rowan and H. Spielmann, Eds.), pp. 617–620. Amsterdam: Elsevier.

Hart, L.A., Wood, M.W., and Weng, H.-Y. (2005). Mainstreaming alternatives in veterinary medical education: Resource development and curricular reform. *Journal of Veterinary Medical Education* 32:473–480.

———. (2006). Three barriers obstructing mainstreaming alternatives in k-12 education. *ALTEX* 23:13–16.

Hatano, G., and Inagaki, K. (1997). Qualitative changes in intuitive biology. *European Journal of Psychology of Education* 12:111–130.

Haury, D.L. (2001). The state of state proficiency testing in science. *ERIC Digest ED465544.* Web site accessed on January 15, 2007, http://www.ericdigests.org/

Hediger, H. (1950/1964). *Wild animals in captivity.* New York City: Dover Publications.

———. (1969). *Man and animal in the zoo: Zoo biology.* New York City: Delacorte Press.

Herbart, J.F. (1835/1901). *Outlines of educational doctrine* (C. DeGarmo, Ed.; A. Lange, Trans.). New York: Macmillan.

Hillar, M. (1997). *The case of Michael Servetus (1911–1553)—The turning point in the struggle for freedom of conscience. Texts and Studies in Religion*, Vol. 74. Lewiston: The Edwin Mellen Press.

Holden, C. (1987). Apples, frogs, and animal rights. *Science* 238:1345.

———. (2003a). Anatomy classes face gross shortage. *Science* 299:1309.

———. (2003b). Texas resolves war over biology texts. *Science* 302:1130.

———. (2004). Disappointing news from Ohio. *Science* 303:1761.

Home Office. (2007). *Scientific procedures on living animals.* London, Great Britain. Web site accessed on July 11, 2007, http://www.homeoffice.gov.uk/rds/scientific1.html

Home Training Tools. (2007). Home science tools: The gateway to discovery. Web site accessed on February 12, 2007, http://www.hometrainingtools.com/catalog/life-science-biology/dissection/cat_preserved-specimens.html?products_begin=0

Hovey, A., Hazelwood, C., and Svedkawskaite, A. (2005). Critical issue: Science education in the era of No Child Left Behind—History, benchmarks, and standards. North Central Regional Educational Laboratory. Web site accessed on January 15, 2007, http://www.ncrel.org/sdrs/areas/issues/content/cntareas/science/sc600.pdf

Humane Society of the United States. (2007a). Humane education loan program (HELP). Web site accessed on July 5, 2007, http://www.hsus.org/animals_in_research/animals_in_education/humane_education_loan_program_help/

———. (2007b). Dissection campaign packet. Web site accessed on July 5, 2007, http://www.hsus.org/animals_in_research/animals_in_education/dissection_campaign_packet.html

————. (2007c). Animals in education. Web site accessed on July 2, 2007, http://www.hsus.org/animals_in_research/animals_in_education/

————. (2007d). Dissection laws. Web site accessed on July 5, 2007, http://www.hsus.org/animals_in_research/animals_in_education/dissection_laws.html

Hurren, E.T. (2004). A pauper dead-house: The expansion of the Cambridge Anatomical Teaching School under the late-Victorian Poor Law, 1870–1914. *Medical History* 48:69–94.

Huxley, T.H. (1854/1902). On the educational value of the natural history sciences. In: *Science and education.* New York: P. F. Collier & Son.

————. (1869/1902). Scientific education. In: *Science and education.* New York: P. F. Collier & Son.

————. (1876/1902). On the study of biology. In: *Science and education.* New York: P. F. Collier & Son.

————. (1877/1902). On elementary education in physiology. In: *Science and education.* New York: P. F. Collier & Son.

Inagaki, K., and Hatano, G. (2002). *Young children's naïve thinking about the biological world.* New York: Psychology Press.

Institute of Laboratory Animal Research. (1996). *Guide for the care and use of laboratory animals.* Washington, DC: National Research Council, National Academy Press.

Institute of Laboratory Animal Resources. (2006). *Principles and guidelines for the use of animals in precollege education.* Washington, DC: National Research Council. www.national-academies.org/ilar

International Association of Human-Animal Interaction Organizations. (2001). The IAHAIO Rio declaration on pets in schools. Web site accessed on July 8, 2007, http://www.iahaio.org/html/rio_declaration.htm

InterNICHE. (2007). International Network for Humane Education. Web site accessed on July 6, 2007, http://www.interniche.org/

Ireton, S.W. (1996). NSTA's Pathways Project: Using the National Science Education Standards. In: *Issues in science education* (J. Rhoton and P. Bowers, Eds.), pp. 46–48. Arlington, VA: National Science Teachers Association.

Jaffe, B. (2003). William Hogarth and eighteenth century English law relating to capital punishment. *Law and Literature* 15:267–278.

Janda, J.M. (2002). On the disappearance of biology labs. *The Scientist* 16(13):11.

Janofsky, M. (2005, December 8). Report says states aim low in science classes. *New York Times.* Web site accessed on January 15, 2007, http://www.nytimes.com/auth/login?URI=http://www.nytimes.com/2005/12/08/education/08schools.html

Johns Hopkins University Center for Alternatives to Animal Testing. (Undated). Dissection. Baltimore, MD: Johns Hopkins Center for Alternatives to Animal Testing.

Jukes, N., and Chiuia, M. (2003). *From guinea pig to computer mouse: Alternative methods for a progressive, humane education,* 2nd ed. Leicester, England: InterNICHE.

K-12 Alliance. (2006). A guide to selecting & purchasing science instructional materials. Web site accessed on July 10, 2007, http://www.k12alliance.net/science_materials.html

Kaiser Family Foundation. (2007). Daily women's health policy. Web site accessed on June 20, 2007, www.kaisernetwork.org

Kariuki, P., and Paulson, R. (2001). The Effects of computer animated dissection versus preserved animal dissection on the student achievement in a high school biology class. Unpublished paper, presented at the Annual Meeting of the Mid-South Educational Research Association, Little Rock, Arkansas.

Kellert, S.R. (1985). Attitudes toward animals: Age-related development among children. *Journal of Environmental Education* 16:29–39.

Kellert, S.R. (1996). *The value of life.* New York: Island Press.

Kellert, S.R., and Westervelt, M.O. (1983). *Children's attitudes, knowledge, and behavior toward animals.* Government Printing Office Report No. 024-010-00641-2. Washington, DC.

Kidd, A.H., and Kidd, R.M. (1987). Reactions of infants and toddlers to live and toy animals. *Psychological Reports* 61:455–464.

———. (1990). Factors in children's attitudes toward pets. *Psychological Reports* 66:775–786.

———. (1997). Changes in the behavior of pet owners across generations. *Psychological Reports* 80:195–202.

Kiesler, S., Sproull, L., and Eccles, J. (1985). Poolhalls, chips, and wargames: Women in the culture of computing. *Psychology of Women Quarterly* 9:451–462.

King, L. (2007). The standards complaint. *USA Today* June 7:11D.

King, L.A., Ross, C.L., Stephens, M.L., and Rowan, A.N. (2004). Biology teachers' attitudes to dissection and alternatives. *ATLA* 32(Suppl. 1):475–484.

Kinzie, M., Larsen, V., Burch, J., and Baker, S. (1996). Frog dissection via the World Wide Web: Implications for widespread delivery of instruction. *Educational Technology Research and Development* 44:59–69.

Kinzie, M.B., Strauss, R., and Foss, J. (1993). The effects of an interactive dissection simulation on the performance and achievement of high school biology students. *Journal of Research in Science Teaching* 30:989–1000.

Kirsch, I.S., Jungeblut, A., Jenkins, L., and Kolstad, A. (1993). *Adult literacy in America: A first look at the findings of the National Adult Literacy Survey.* #NCES 93275. Institute of Education Sciences, National Center for Education Statistics, U.S. Department of Education. Web site accessed on August 12, 2007, http://nces.ed.gov/pubsearch/pubsinfo.asp?pubid=93275

Kisling, V.N., Jr. (Ed.) (2001). *Zoo and aquarium history: Ancient animal collections to zoological gardens.* Boca Raton, FL: CRC Press.

Klestinec, C. (2004). A history of anatomy theaters in sixteenth century Padua. *Journal of the History of Medicine* 59(3):375–412.

Kline, A.D. (1995). We should allow dissection of animals. *Journal of Agricultural and Environmental Ethics* 8:190–197.

Koballa, T.R., Jr. (1995). Children's attitudes toward learning science. In: *Learning science in the schools: Research reforming practice* (S.M. Glynn and R. Duit, Eds.), pp. 59–84. Mahwah, NJ: Lawrence Erlbaum Associates.

Kotrschal, K., and Ortbauer, B. (2003). Behavioral effects of the presence of a dog in a classroom. *Anthrozoos* 16:147-159.

Krueger, A., and Sutton, J. (Eds.) (2001). *ED thoughts: What we know about science teaching and learning.* Aurora, CO: Mid-Continent Research for Education and Learning.

Kuhn, T.S. (1962). *The structure of scientific revolutions.* Chicago: University of Chicago Press.

Langley, G.R. (1991). Animals in science education—ethics and alternatives. *Journal of Biological Education* 25:274–279.

LaRosa, R. (2006). Waking up a nation gone drowsy on science. *San Francisco Chronicle* February 9:B9.

Lawrence Hall of Science. (2007). FOSS—Full Option Science System. Lawrence Hall of Science, University of California, Berkeley. Web site accessed on May 23, 2007, http://www.lhs.berkeley.edu/foss/index.html

Lemonick, M.D. (2006, February 13). Is America flunking science? *Time.*

Library of Congress. (1990). The crisis in science education. Web site accessed on May 23, 2007, http://www.loc.gov/rr/scitech/tracer-bullets/crisis-edutb.html

Lieb, M.F. (1985). Dissection: A valuable motivational tool or a trauma to the high school student? Unpublished Thesis, Master of Education, National College of Education, Evanston, Illinois.

Lock, R. (1994). Dissection as an instructional technique in secondary science: Comment on Bowd. *Society & Animals* 2:67–73.

Lock, R., and Millett, K. (1992). Using animals in education and research: Student experience, knowledge and implications for teaching in the National Science Curriculum. *School Science Review* 74:115–123.

Lowery, L. F. (1992). *The scientific thinking processes.* Berkeley, CA: Full Option Science System (FOSS), Lawrence Hall of Science.

Lucile Packard Foundation for Children's Health. (2006). Facts by topic: High school graduates completing college preparatory courses: 1995–2005. Web site accessed on September 15, 2006, http://www.kidsdata.org/topictrends.jsp?i=1&t=28&ra=2_1

Maehle, A.-H. (1993). The ethical discourse on animal experimentation, 1650–1900. *Clio Medica* 24:203–251.

Magruder, K. (2007). 16th Century life sciences. In: *History of Science, the University of Oklahoma.* Website accessed on August 9, 2007, http://hsci.cas.ou.edu/exhibits/exhibit.php?exbgrp=-999&exbid=48&e . . .

Marks, A. (2005, April 19). States hit back on school reform law—Connecticut, Utah, and Texas are either refusing to adopt all of the No Child Left Behind Act or suing the US to block it. *The Christian Science Monitor.*

Marshall, T. (1995). *Murdering to dissect: Grave-robbing, Frankenstein, and the anatomy literature.* Manchester, England: Manchester University Press.

Marszalek, C.S., and Lockard, J. (1999). Which way to jump: Conventional frog dissection, CD-tutorial, or Microworld? Proceedings of Selected Research and Development Papers presented at the National Convention of the Association for Educational Communication and Technology, Houston, TX.

Matthews, M.R. (Ed.) (1998). *Constructivism in science education: A philosophical examination.* Dordrecht, the Netherlands: Kluwer Academic Publishers.

———. (2000). *Time for science education: How teaching the history and philosophy of pendulum motion can contribute to science literacy.* New York: Kluwer Academic/Plenum Publishers.

McCabe, K. (1990). Beyond cruelty. *The Washingtonian* 25(5):72.

McCollum, T.L. (1987). The effect of animal dissections on student acquisition of knowledge of and attitudes toward the animals dissected. Unpublished doctoral dissertation, University of Cincinnati.

McInerney, J. (1993). Animals in education: Are we prisoners of false sentiment? *American Biology Teacher* 55:276–280.

McNicholas, J., and Collis, G.M. (2001). Children's representations of pets in their social networks. *Child: Care, Health & Development* 27:279–294.

Melson, G.F. (1988). Availability of and involvement with pets by children: Determinants and correlates. *Anthrozoos* 2:45–52.

Mervis, J. (2003). Report asks colleges to plug a leaky people pipeline. *Science* 300:1353.

Messybeast.com Cat Resource Archive. (2007). Cat theft for classroom dissection studies: "Catching the cat thieves." Web site accessed on August 28, 2006, http://messybeast.com/cat-dissect.html

Mintzes, J.J., and Wandersee, J.H. (1997a). Reform and innovation in science teaching: A human constructivist view. In: *Teaching science for understanding: A human constructivist view* (J.L. Mintzes, J.H. Wandersee, and J.D. Novak, Eds.), pp. 29–58. San Diego: Academic Press.

———. (1997b). Research in science teaching and learning: A human constructivist view. In: *Teaching science for understanding: A human constructivist view* (J.L. Mintzes, J.H. Wandersee, and J.D. Novak, Eds.), pp. 59–92. San Diego: Academic Press.

Montessori, M. (1912). *The Montessori method; scientific pedagogy as applied to child education in "The children's houses".* New York: Frederick A. Stokes Company.

Moscoso, J. (1998). Monsters as evidence: The uses of the abnormal body during the early eighteenth century. *Journal of the History of Biology* 31:355–382.

Mullins, M.E. (2006). Gender gaps. *USA Today* October 13:11A.

Murray, C. (2007a, January 16). On education: Intelligence in the classroom. *Wall Street Journal* A21.

———. (2007b, January 17). On education: What's wrong with vocational school? *Wall Street Journal* 17: A19.

———. (2007c, January 18). On education: Aztecs vs. Greeks. *Wall Street Journal* A17.

Muzslay, L. (2006). Press release: Restrictive textbook adoption policy adds fuel to math, science teacher shortage. Web site accessed on February 12, 2007, http://www.keypress.com/pressroom/CSU_Summit.html

Nabi, R.L. (2002). The theoretical versus the lay meaning of disgust: Implications for emotion research. *Cognition and Emotion* 16:695–703.

NARAL Pro-Choice America Foundation (National Association for the Repeal of Abortion Laws). (2007). Americans support responsible sex education. Web site accessed on June 20, 2007, http://www.PROCHOICEAmerica.org

NASCO. (2007). NASCO online catalogs. Web site accessed on February 12, 2007, http://www.eNasco.com/science

National Academies Press. (1996/2007). *National science education standards.* Washington, DC: National Academy Press. Web site accessed on July 6, 2007, http://www.nap.edu/readingroom/books/nses/html/overview.html

National Assessment of Educational Progress (2007). NAEP and No Child Left Behind. Web site accessed on May 22, 2007, http://nces.edu.gov/nationsreportcard/nclb.asp

National Association for Biomedical Research. (2007). Dissection in education. Web site accessed on July 12, 2007, http://www.nabr.org/AnimalLaw/Education/Dissection.htm

National Association for Humane and Environmental Education. (2007). NA-HEE: Alternatives to dissection. Web site accessed on February 12, 2007, http://www.nahee.org/

National Association for the Advancement of Humane Education. (1985). The living science: A human approach to the study of animals in elementary and secondary biology—A brochure for teachers. East Haddam, CT: The National Association for the Advancement of Humane Education.

National Association of Biology Teachers. (2003). The use of animals in biology education. Web site accessed on February 12, 2007, http://www.nabt.org/sub/position_statements/animals.asp

———. (2007). Resource center. Web site accessed on February 12, 2007, http://www.nabt.org/

National Association of State Boards of Education. (2007). State-level school health policies: State-by-state alcohol, tobacco, and drug use education. Web site accessed on May 23, 2007, http://www.nasbe.org/healthyschools

National Center for Education Statistics. (2006). The nation's report card: Science 2005. Web site accessed on May 22, 2007, http://nces.edu.gov/pubsearch/pubsinfo.asp?pubid=2006466

National Commission on Excellence in Education. (1983). A nation at risk: The imperative for educational reform. Web site accessed on May 23, 2007, http://www.ed/gov/pubs/NatAtRisk?/index.html

National Education Association. (1893). *Report of the committee on secondary school studies.* Washington, DC: US Government Printing Office.

———. (1918). *Cardinal principles of secondary education: A report of the commission on the reorganization of secondary education.* (U.S. Bureau of Education, Bulletin No. 35). Washington, DC: U.S. Government Printing Office.

———. (2006). ESEA: It's time for a change! NEA's positive agenda for the ESEA reauthorization. Web site accessed on January 15, 2007, http://www.nea.org/lac/esea/images/posagenda.pdf

National Library of Medicine. (2007). The visible human. Web site accessed on February 12, 2007, www.nlm.nih.gov/research/visible/visible_human.html

National Research Council. (1996). *National science education standards.* Washington, DC: National Academy Press.

National School Boards Association. (2007). NSBA's school health programs: State-level school health policies: State-by-state alcohol, tobacco, and drug use education. Web site accessed on May 24, 2007, http://www.nasbe.org/healthyschools

National Science Teachers Association. (1996). *Pathways to the science standards: Guidelines from moving the vision into practice.* Arlington, VA: National Science Teachers Association.

———. (2002a). "No Child Left Behind" Act offers new professional development opportunities for science educators. Web site accessed on February 12, 2007, http://www3.nsta.org/main/news/stories/education_story.php?news_story_ID=47257

———. (2002b). Federal funding for K-12 science and math education is woefully inadequate, say NSTA and partner groups. Web site accessed on January 15, 2007, http://www.nsta.org/pressroom&news_story_ID=47188&print=y

———. (2005). Responsible use of live animals and dissection in the science classroom. Web site accessed on February 12, 2007, http://www.nsta.org/positionstatement&psid=44

———. (2007). NSTA. National Science Teachers Association. Website accessed on February 12, 2007, http://www.nsta.org/

NORINA (Norwegian Reference Centre for Laboratory Animal Science and Alternatives). (2007). NORINA: A Norwegian inventory of audiovisuals. Web site accessed on February 12, 2007, http://oslovet.veths.no/NORINA

Novak, J.D. (1997). The pursuit of a dream: Education can be improved. In: *Teaching science for understanding: A human constructivist view* (J.J. Mintzes, J.H. Wandersee, and J.D. Novak, Eds.), pp. 3–28. San Diego, CA: Academic Press.

O'Malley, C.D. (1964). *Andreas Vesalius of Brussels: 1514–1564*. Berkeley: University of California Press.

Organization for Economic Cooperation and Development (OECD). (2005). Education at a glance 2005. Web site accessed on October 22, 2005, http://www.oecd.org/edu

Orlans, F.B. (1991). Use of animals in education: Policy and practice in the United States. *Journal of Biological Education* 25:27-32.

———. (1993). *In the name of science: Issues in responsible animal experimentation*. New York: Oxford University.

———. (1998). History and ethical regulation of animal experimentation: An international perspective. In: *A companion to bioethics* (H. Kuhse and P. Singer, Eds.), pp. 399–410. Oxford, UK: Blackwell Publishers.

———. (2000). An ethical rationale for why students should not be permitted to harm or kill animals. In: *Progress in the reduction, refinement and replacement of animal experimentation* (M. Balls, A.M. van Zeller, and M. Halder, Eds.), pp. 1323–1331. Amsterdam: Elsevier.

Ormerod, M.C., and Duckworth, D. (1975). *Pupils' attitudes to science: A review of research*. Berkshire, England: NFER Publishing.

Patronek, G.J., and Rauch, A. (2007). Systematic review of comparative studies examining alternatives to the harmful use of animals in biomedical education. *Journal of the American Veterinary Medical Association* 230:37–43.

Paul, E.S., and Podberscek, A.L. (2000). Veterinary education and students' attitudes towards animal welfare. *Veterinary Record* 146:269–272.

Pavletic, M.M., Schwartz, A., Berg, J., and Knapp, D. (1994). An assessment of the outcome of the alternative medical and surgical laboratory program at Tufts University. *Journal of the American Veterinary Medical Association* 205:97–100.

People for the Ethical Treatment of Animals (PETA). (2007). PETA's history: Compassion in action. Web site accessed on July 14, 2007, http://www.peta.org/factsheet/files/FactsheetDisplay.asp?ID=107

Performing Animal Welfare Society. (2007). Performing Animal Welfare Society. Web site accessed on July 8, 2007, http://www.pawsweb.org/site/homepage.htm

Physicians Committee for Responsible Medicine (PCRM). (2007). Cost analysis of dissection versus nonanimal teaching methods. Web site accessed on July 9, 2007, http://www.pcrm.org/resch/anexp/cost_analysis.html

Pifer, L.K. (1994). Adolescents and animal research: Stable attitudes or ephemeral opinions? *Public Understanding of Science* 3:291–307.

Plous, S. (1996). Attitudes toward the use of animals in psychological research and education: Results from a national survey of psychologists. *American Psychologist* 51:1167–1180.

Project Reality. (2001). A.C. Green's game plan abstinence program. Core-Alliance Group, Inc. Web site accessed on June 17, 2007, http://projectreality.org/products/index.php?id=8

Ra'anan, A.W. (2005). The evolving role of animal laboratories in physiology instruction. *Advances in Physiology Education* 29:144–150.

Raloff, J. (2001). Where's the book? Science education is redefining texts. *Science News* 159:186–188.

Ralph, C.L. (1996). The illusion of the instructional biology laboratory. *American Biology Teacher* 58:142–146.

Ratzan, R.M. (2004). Essay review: *A traffic of dead bodies*. *Perspectives in Biology and Medicine* 47(2):290–299.

Research!America. (2007). America speaks. Web site accessed on June 30 2007, http://www.researchamerica.org/publications/AmericaSpeaks/AmericaSpeaks V8.pdf

Richardson, R. (2000). *Death, dissection and the destitute*. Chicago: University of Chicago Press.

Richardson, R. (2006). Human dissection and organ donation: A historical and social background. *Mortality* 11:151–165.

Ricketts, M. (2006). *Gallery of great painters: Rembrandt*. Lisse, the Netherlands: Rebo International.

Rosenberg, H., and Fitch, W. (1998). How to reduce the level of formaldehyde in the zoology lab. In: *Tested studies for laboratory teaching* (S. J. Karcher, Ed.), *Proceedings of the 19th Workshop/Conference of the Association for Biology Laboratory Education*, 19:357–360. Web site accessed on October 19, 2007, http://www.zoo.utoronto.ca/able/volumes/vol-19/mini.12.rosenberg.pdf

Rudner, L.M. (1999). Scholastic achievement and demographic characteristics of home school students in 1998. *Education Policy Analysis Archives* 7(8). Web site accessed October 19, 2007, http://epaa.asu.edu/epaa/v7n8/

Rudolph, J.L. (2002). *Scientists in the classroom: The cold war reconstruction of American science*. New York City: Palgrave.

Sachs, O. (2001). *Uncle Tungsten: Memories of a chemical boyhood*. New York: Vintage Books.

San Diego Unified School District. (2005). Health education, pupil services, and parents' or students' rights requiring annual notification. Web site accessed on June 22, 2007, http://www.dmusd.org/district/files/openFile.aspx?fileID=1792

San Francisco Chronicle. (2007, May 27). Our diploma-less students, p. E4.

Sapontzis, S.F. (1995). We should not allow dissection of animals. *Journal of Agricultural and Environmental Ethics* 8:181–189.

Sappol, M. (2002). *A traffic of dead bodies: Anatomy and embodied social identity in nineteenth-century America*. Princeton, NJ: Princeton University Press.

———. (2003). The anatomical mission to Burma. *Science* 302:232–233.

Sauer, U.G., Spielmann, H., and Rusche, B. (2005). Fourth EU report on the statistics on the number of animals used for scientific purposes in 2002—Trends, problems, conclusions. *ALTEX* 22:59–67.

Sawday, J. (1995). *The body emblazoned: Dissection and the human body in Renaissance culture.* London: Routledge.

Science Education for Public Understanding Program (SEPUP). (2005). *Science & life issues.* Lawrence Hall of Science, University of California at Berkeley, Ronkonkoma, NY: Lab-Aids, Inc.

Sejnowski, T.J. (2003). Tap into science 24-7. *Science* 301:601.

Shanks, N., and Dawkins, R. (2004). *God, the devil, and Darwin: A critique of intelligent design theory.* New York: Oxford University Press.

SIECUS. (2005). SIECUS state profiles: A portrait of sexuality education and abstinence-only-until-marriage programs in the states released today. Web site accessed on June 15, 2007, http://www.siecus.org/media/press/press0095.html

Singer, S.R., Hilton, M.L., and Schweingruber, H.A. (Eds.) 2005. *America's lab report: Investigations in high school science.* Washington, DC. Committee on High School Science Laboratories: Role and Vision, National Research Council. Web site accessed on June 17, 2007, http://www.nap.edu/catalog/11311.html

Sizer, T.R. (1964). *Secondary schools at the turn of the century.* New Haven, CT: Yale University Press.

Skinner, B.F. (1938). *The behavior of organisms.* New York: Appleton Century Crofts.

Skinner, E.A., and Belmont, M.J. (1993). Motivation in the classroom: Reciprocal effects of teacher behaviour and student engagement across the school year. *Journal of Educational Psychology* 85:571–581.

Smith, A. (2007). The Norwegian Reference Centre for Laboratory Animal Science (NORINA). Web site accessed on October 22, 2007, http://oslovet.veths. no/NORINA/

Smith, O. (2004). Education can be fun. *Science* 305:1108–1109.

Smith, W. (1994). Use of animals and animal organs in schools: Practice and attitudes of teachers. *Journal of Biological Education* 28:111–118.

Solot, D., and Arluke, A. (1997). Learning the scientist's role: Animal dissection in middle school. *Journal of Contemporary Ethnography* 26:28–54.

Spence, T.F. and Zuckerman, S. (1967). *Teaching and display techniques in anatomy and zoology.* Oxford, England: Pergamon Press.

Spencer, H. (1864). *Education: Intellectual, moral, and physical.* New York: Appleton.

Squire, K., and Jenkins, H. (2003). Harnessing the power of games in education. The Institute for the Advancement of Emerging Technology in Education. Web site accessed on July 6, 2007, http://www.Edvantia.org

Stello, C. (2006, October 13). Speaker: Public schools are separate, unequal. *Davis Enterprise* A3.

Stokes, W.S. (1997). Animal use alternatives in research and testing: Obligation and opportunity. *Lab Animal* 26(3):28–32.

Stokes, W.S., and Jensen, D.J.B. (1995). Guidelines for institutional animal care and use committees: Consideration of alternatives. *Contemporary Topics* 34:51–60.

Storm, B. (2007). Valley Oak Science. Web sites accessed on July 2007, http://www. djusd.k12.ca.us/valleyoak/bstorm/ and http://www.djusd.k12.ca.us/valleyoak/ bstorm/biosafarishome.htm

Strauss, R.T., and Kinzie, M.B. (1994). Student achievement and attitudes in a pilot study comparing an interactive videodisc simulation to conventional dissection. *American Biology Teacher* 56:398–402.

Stuart, M.D., and Henry, R.W. (2002). Plastinated specimens can improve the conceptual quality of biology labs. *American Biology Teacher* 64:130–134.

Sund, R.B., and Trowbridge, L.W. (1967). *Teaching science by inquiry in the secondary school.* Columbus, OH: Charles E. Merrill Books.

Sweitzer, J.S. (1996). Slice of life: Virtual dissection is still a cut below the real thing. *The Sciences* 36(2):41–43.

Tamir, P., and Shcurr, Y. (1997). Back to living animals: An extracurricular course for fifth-grade pupils. *Journal of Biological Education* 31:300–304.

Taylor, D. (1998). *Family literacy: Young children learning to read and write.* Portsmouth, NH: Heinemann.

Texley, J. (1992). Doing without dissection. *American School Board Journal* 179(1):24–265

Toppo, G. (2007, January 8). How Bush education law has changed our schools. *USA Today* 1–2A.

Torrance Unified School District. (2005). HIV/AIDS prevention; Personal and public safety, accident prevention and health. California School Boards Association. Web site accessed on June 22, 2007, http://www.tusd.org/Pages/supt/BdPolicy/6142_92-6146_5.pdf

Treagust, D.F., Duit, R., and Fraser, B.J. (1996). Overview: research on students' preinstructional conceptions—the driving force for improving teaching and learning in science and mathematics. In: *Improving teaching and learning in science and mathematics* (D.F. Treagust, R. Duit, and B.J. Fraser, Eds.), pp. 1–14. New York: Teachers College Press, Columbia University.

Trends in International Mathematics and Science Study (TIMSS) (Martin, M.O., Mullis, I.V.S., Gonzalez, E.J., and Chrostowski, S.J). (2004). *Findings from IEA's Trends in International Mathematics and Science Study at the fourth and eighth grades.* Chestnut Hill, MA: TIMSS & PIRLS International Study Center, Boston College. Web site accessed on October 22, 2005, http://timss.bc.edu/timss2003i/scienceD.html

Trowbridge, J.E. and Mintzes, J.J. (1985). Students' alternative conceptions of animal classification. *School Science and Mathematics* 85:304–316.

Tsuzuki, M., Asada, Y., Akiyama, S., Macer, N., and Macer, D. (1998). Animal experiments and bioethics in high schools in Australia, Japan, and New Zealand. *Journal of Biological Education* 32:119–126.

Tunnicliffe, S.D. (1996a). A comparison of conversations of primary school groups at animated, preserved, and live animal specimens. *Journal of Biological Education* 30:195–206.

———. (1996b). Conversations within primary school parties visiting animal specimens in a museum and zoo. *Journal of Biological Education* 30:130–141.

———. (1996c). Expressed attitudes of primary school and family groups to animal exhibits. *ISAZ The Newsletter* 12:7–12.

———. (1996d). The relationship between pupils' age and the content of conversations generated at three types of animal exhibits. *Research in Science Education* 26:461–480.

———. (1997a). The effect of the presence of two adults—Chaperones or teachers—On the content of the conversations of primary school groups during school visits to a natural history museum. *Journal of Elementary Science Education* 9:49–64.

———. (1997b). Is it zoology? Zoo visits for primary school groups. *Biologi Italiani* 27:34–39.

———. (1998a). Boy talk/girl talk: Is it the same at animal exhibits? *International Journal of Science Education* 20:795–811.

———. (1998b). The content of conversations generated by school children viewing live animals as exhibits and on a farm. *Journal of Elementary Science Education* 10:1–17.

Tunnicliffe, S.D., Lucas, A.M., and Osborne, J. (1997). School visits to zoos and museums: A missed educational opportunity? *International Journal of Science Education* 19:1039–1056.

UC Center for Animal Alternatives. (2007). Tutorial for educational alternatives sites. Web site accessed on February 12, 2007, http://www.vetmed.ucdavis.edu/Animal_Alternatives/altsearch.htm

University of California Office of the President. (2007a). National health education standards. California Physical Education-Health Project, University of California. Web site accessed on July 18, 2007, http://csmp.ucop.edu/cpehp/standards/hpnational.html

———. (2007b). The California challenge standards in health education; Questions. California Physical Education-Health Project, University of California. Web site accessed on July 7, 2007, http://csmp.ucop.edu/cpehp/standards/hpchallenge.html; http://www.cde.ca.gov/ci/he/he/

University of California, Davis, School of Veterinary Medicine. (2007). The virtual heart. Web site accessed on June 15, 2007, www.calf.vetmed.ucdavis.edu

Urban Cat Project. (2006). Urban Cat Project: Feline rescue, adoption, advocacy. Web site accessed on June 11, 2006, http://www.theurbancatproject.org/exploit.html

U.S. Department of Agriculture. Animal Welfare Act 7USC 2131-2159, adopted 1966, amended 2002. Public Law 89-544, 1966, as amended (P.L. 91-579, P.L. 94-279, and P.L. 99-198) 7 U.S.C. 2131 et seq. Implementing regulations are published in the Code of Federal Regulations (CFR), Title 9, Chapter 1, Subchapter A, Parts 1, 2, and 3. Web site accessed on February 12, 2007, http://www.aphis.usda.gov/ac/awa.html

U.S. Department of Agriculture. (1997). *Policy 11: Painful procedures.* Animal resource guide. Web site accessed on August 12, 2007, http://www.aphis.usda.gov/ac/policy/policy11.pdf

U.S. Department of Education. (Institute of Education Sciences. National Assessment of Educational Progress). (2002a). The nation's report card: Science 2000. In: *NCES 2003–453* (C.Y. O'Sullivan, M.A. Lauko, W.S. Grigg, J. Qian, and J. Zhang, Eds.). Washington, DC: National Center for Education Statistics. Web site accessed on October 22, 2005, http://nces.ed.gov/nationsreportcard/pubs/main2000/2003453.asp

———. (2002b). No child left behind. Web site accessed on July 7, 2007, http://www.ed.gov/policy/elsec/leg/esea02/107-110.pdf

———. (2004). New No Child Left Behind flexibility: Highly qualified teachers. Web site accessed on January 15, 2007, http://www.ed.gov/nclb/methods/teachers/hqtflexibility.pdf

U.S. Department of Education. (Institute of Education Sciences. National Assessment of Educational Progress). (2007). The nation's report card. Web site accessed on July 6, 2007, http://nces.ed.gov/nationsreportcard/science/results/

Van der Valk, J., Dewhurst, D., Hughes, I., Atkinson, J., Balcombe, J., Braun, H., Gabrielson, K., Gruber, F., Miles, J., Nab, J., Nardi, J., vanWilgenburg, H., Zinko, U., and Zurlo, J. (1999). Alternatives to the use of animals in higher education: The report and recommendations of ECVAM Workshop 33. *ATLA* 27:39–52.

Vergano, D. (2006, February 9). U.S. could fall behind in global "brain race." *USA Today* 1D–2D.

Vesalius, A. (1543/1964). *De humani corporis fabrica libri septem*, p. 548. Bruxelles: Culture et Civilization.

Von Hagens, G., and Whalley, A. (2002, 3rd printing). BODY WORLDS: The anatomical exhibition of real human bodies. Institute for Plastination, Druck, Leimen, Germany. Web site accessed on February 12, 2007, http://www.bodyworlds.com

WAIT Training. (2007). WAIT training. Web site accessed on June 22, 2007, http://www.waittraining.com/

WARD'S Natural Science. (2007). WARD's natural science. Web site accessed on October 26, 2007, http://www.wardsci.com/

Weng, H.-Y., Wood, M.W., and Hart, L.A. (2004). Educators. Web site accessed on June 30, 2007, http://www.vetmed.ucdavis.edu/Animal_Alternatives/educator.htm

Wenning, R., Herdman, P.A., Smith, N., McMahon, N., and Washington, K. (2003). No child left behind: Testing, reporting and accountability. *ERIC Digest ED480994*. Web site accessed on January 15, 2007, http://www.ericdigests.org/

Yager, R.E. (1996). Science teacher preparation as a part of systemic reform in the United States. In: *Issues in science education* (J. Rhoton and P. Bowers, Eds.), pp. 24–33. Arlington, VA: National Science Teachers Association.

Zasloff, R.L., and Hart, L.A. (1997). Adapting animal alternatives from veterinary medical education to precollege education. In: *Animal alternatives, welfare and ethics* (L.F.M. van Zutphen and M. Balls, Eds.), pp. 445–447. Amsterdam: Elsevier.

Zasloff, R.L., Hart, L.A., and De Armond, H. (1999). Animals in elementary school education in California. *Journal of Applied Animal Welfare Science* 2:347–357.

Zasloff, R.L., Hart, L.A., and Weiss, J.M. (2004). Dog training as a violence prevention tool for at-risk adolescents. *Anthrozoos* 16:352–359.

Zirkel, J.B., and Zirkel, P.A. (1997). Technological alternatives to actual dissection in anatomy instruction: A review of the research. *Educational Technology* 10(3):52–56.

Index

AAAS. *See* American Association for the Advancement of Science

AALAS. *See* American Association for Laboratory Animal Science

Achieve (2007), 66

Active learning, 90–91

ACT test, 54, 69

Adkins, J., 14, 121, 153, 155

Alcohol use, 96–97

Alcott, William Andrus, 32, 34, 36, 37

Alternatives, 50–51, 131, 163; availability of, 50–51, 153; barriers against, 166; databases for teaching, 185–91; dissection v., 147; funding for, 51; medical school use of, 176–78; problems of introducing, 49; students electing, 51; veterinary education introduction of, 175–78. *See also* Teaching resources

AltWeb, 191

American Alliance for Health, Physical Education, Recreation and Dance, 82

American Association for Laboratory Animal Science (AALAS), 108, 123

American Association for the Advancement of Science (AAAS), 54, 78, 90

American Association of Medical Colleges (AAMC), survey by, 88–89

American Physiological Society, 89, 153

Anatomy Act in 1832, 32

"The Anatomy Lesson of Dr. Jan Deyman" (Rembrandt), 29

"The Anatomy Lesson of Dr. Nicolaes Tulp" (Rembrandt), 29

Anatomy, study of, 11, 18, 20, 24, 25, 26, 28, 139; body parts for, prosected, 167; cardiovascular, 27; computer simulation for, 90, 91, 131, 147, 163; democratization of, 139; history/importance of, 19–20; human, animal use in, 167, 183; schools of, London, 34

Animal(s): children's exposure to, 117–18, 134–45, 155; dissected, yearly number of, 153; early view of, 19; emotional experiences regarding, 17, 25, 46, 134–45; fear of, 135; knowledge of, dissection yielding, 146–47; literature containing, children's, 117; pain/distress of, legislation on, 108; as pets, 117, 122, 134, 146;

Animal(s) (*cont.*)
  purpose-bred, 50; sanctuaries for,
  158; species of, most commonly
  dissected, 120–21, 153–55, 159–60;
  as surrogates, 18, 183; vertebrate,
  153–54; whole v. body parts,
  138–39. *See also* Animal use
  Cadavers Specimens
Animalearn, 8, 100, 124, 148, 153, 181,
  196
Animal use, 17–18, 120–21, 167,
  174–76, 178; anatomy study,
  human, 167, 183; classroom, 102–8,
  117, 155, 159–63; consumptive, 175;
  contention on, 3–4, 39, 61, 144;
  educational level and, 2–4, 17, 51,
  143, 144, 153–55; guidelines for,
  108–9; informal education, 45–46,
  143, 157–59; legislation on, 4, 93,
  102–8; live, 155, 157–59; medical
  schools using, decline in, 88–89;
  statistics on, 153, 154; vivisection,
  18, 24, 129
Animal welfare, 46, 102, 153, 160–61;
  organizations, 103–7, 153, 158,
  160–61, 181–82
Anthropology, 38
Apprehension, 137–42
Aristotle of Macedonia, 18, 19
Arluke, A., 142
Artists, role of, early history and, 20,
  28
Ascione, F.R., 134, 137
Asia, 5
Association of Science and Technology
  Centers Network, 47
Association of Veterinarians for
  Animal Rights (AVAR), 13, 50, 102,
  190
Australia, 14, 121
Ausubel, David, 43, 133, 145
Autopsy, 31
AVAR search engine, 184, 185, 190

"The Babe in the Womb" (DaVinci),
  20, 141
Balcombe, Jonathan, 14, 48, 119, 120,
  178

Banning, 51
Barber-surgeons, 19
Basic skills, 35, 44
Behavior/behaviorism, 42, 96
Benedictus, 27
Biological Sciences Curriculum Study
  (BSCS), 42, 43, 60, 175
Biology, 34, 81; animal/human
  comparison in, 183; cell, 83, 85–87;
  children learning, 10, 145–48; death
  and, 143–45; 1800s, 36–39; grades
  9–12, 85–87; health and, 10, 93–98,
  148–49; interest in, 15, 69;
  intermediate/
  secondary school, 118–20, 155;
  laboratories, 51, 88–89; legislation,
  93–98; life sciences and, grade
  school, 83–87; reforms in, Sputnik
  leading to, 42–43; relevance of, 8–9,
  10; resources for teaching, 166,
  168–74; sex education in context of,
  97; subject matter legislation for,
  95–98; testing in, 69
BioSafaris, 176
Birth deformities, 27, 28
Bodies. *See* Cadavers Human bodies
  Specimens
Body parts, 138–39, 167
Bodysnatching, 32
BODY WORLDS, 13, 24, 25, 139, 162,
  184
Bowd, A.D., 14, 145, 148
Boyle, Robert, 28
Brain, 29
BSCS. *See* Biological Sciences
  Curriculum Study
Budget, 116, 119
Burke and Hare scandal, 31–32

CAAT. *See* Center for Alternatives in
  Animal Testing
Cadavers: criminals used as, hanged,
  20, 30, 31–32; live animals v., use of,
  155, 157–59; plastinated, 13, 24, 28,
  162–63; poses of, 24, 25, 139. *See
  also* Specimens
Cadavers, acquisition of, 50, 159–61;
  history of, 19, 20, 22, 30–34

California, 54, 71–78, 87–88
California Science Teachers
    Association (CSTA), 54, 180
California State Board of Education,
    83, 85, 88
Cambridge anatomical Teaching
    School, 32
Canada, 14
Cardiovascular anatomy, 27
Carlino, A., 19, 20, 22, 28
Carnival season, 22, 27, 31
Carolina Biologicals, 13, 152, 159, 160,
    161–62, 181
Cats, 120–21, 161
Cell biology, 83, 85–87
Center for Alternatives in Animal
    Testing (CAAT), 191
Ceremony, 27–28, 144–45
Children: animal exposure of, 117–18,
    134–45, 155; biology learning of, 10,
    145–48; death understanding of,
    143–45; knowledge of living things
    in, 142–43; schedules of, school,
    9–10, 116
Christianity, 17, 31, 34
Circulatory system, 22
Classrooms: animal use in, 102–8, 117,
    155, 159–63; learning in families v.,
    156; specimen use/dissection in,
    159–63
Clergy, 31
CLIVE. See Computer-Aided Learning
    in Veterinary Education
Colombo, Realdo, 18, 24, 26
Commission on the Reorganization of
    Secondary Education, 41–42
Computer-Aided Learning in
    Veterinary Education (CLIVE), 125
Computer simulations, 90, 91, 131,
    147, 163
Concept, of "living," 142–43. See also
    Learning
Constructivism, 43–45, 133–34
Consultants, 196
Content. See Subject matter
Contention, 60–61; animal use, 3–4,
    39, 61, 144; educational, 61, 62–63,
    66, 95, 96; vivisection, 18, 129

Contention, dissection, 154; animal v.
    human body, 19; Christianity/
    resurrection, 17, 31; in education,
    46–51; human body, 17, 19, 31, 34.
    See also Legislation
Conversational analysis, children, 136
Costs, 91, 161–62
Counseling, 100, 102
Cows, 160
Creationism, 61
Criminals, cadaver use of, 20, 30,
    31–32
CSTA. See California Science Teachers
    Association
Cunningham, A., 18, 20, 27, 28
Curriculum, 54; design, higher
    education, 167, 174; health/biology,
    62, 96; precollege, 178–79;
    standards/
    frameworks not defining, 79;
    teachers designing own, 174–75

Da Foligna, Gentile, 19
Darwin, Charles, 35–36, 38
Databases, alternatives, teaching,
    185–91
Da Vinci, Leonardo, 20
Da Vinci, Leonardo, "The Babe in the
    Womb," 20, 141
Death, 23, 25, 29; children's
    understanding of, 143–45
De Boer, G.E., 36, 38, 60
De Humani Corporis Fabrica (On the
    Construction of the Human Body)
    (Vesalius), 22, 23
Dei Luzzi, Mondino, 19, 20, 21, 139,
    140
Democratization, anatomy, 139
Dewey, John, 41
Disabilities, 69
Disgust, 137–42
Dissection: alternatives to, 49, 50–51,
    90, 91, 122–28, 129–30, 131, 147,
    153, 163, 166; apprehension about,
    137–42; banning of, 51; carnival
    season, 22, 27, 31; ceremony before,
    27–28, 144–45; costs issues
    surrounding, 91, 161–62;

Dissection (*cont.*)
    discretionary use of, 34; excitement
    of early, 27; high school adoption of,
    119; history of, 3, 18–34, 40; on
    humans, 17–18, 19, 31, 34, 166–67,
    183, 191; knowledge from, 19–20,
    21–22, 27, 36, 38; legislation on
    participation in, 101–2; location of,
    26–28; negative memories of, 14; as
    not addressed, 7; obsolescence of,
    116; opting out of, 51, 100, 102,
    147–48; phasing out of, 2, 14, 50, 51;
    popularity of, 28; postmortem, 27;
    precollege classroom, 159–63; in
    precollege v. higher education, 2–4,
    17, 51, 143, 153–55; preparation for,
    142, 144–45; public; purpose of,
    early history and, 19–20, 21–22;
    Rembrandt paintings of, 29;
    statistics on, 153, 154; student
    motivation from, 8–9, 165; studies
    on alternatives v., 147; subjects
    covered by, 116; teachers' view of,
    14, 49; theatres of, 27–28. *See also*
    Animal use; Contention; Studies
Dissectors: barber-surgeons as, 19; Da
    Vinci as, 20, 141; early, 19, 20–21
Diversity, testing as measure of, 65
Dogs, 146; fear of, 135
Drugs, 96–97
Duschl, R.A., 89, 111

Edinburgh, 31–32
Education, 66, 95, 96; basic skills in, 35,
    44; concept formation and, 36, 79,
    142–43; contention over dissection
    in, 46–51; 1800s views of, 36–39;
    goals of, 41–42, 60, 62–63; informal,
    45–46, 143, 157–59; interest/
    disinterest in, 15, 44, 63, 67, 69, 140,
    143; materials in, scientists creating,
    43; museums/learner-centric
    approach to, 45–46; precollege v.
    higher, 2–4, 17, 51, 143, 144,
    153–55; self-directed v. established,
    1900s views of, 39, 41; sex, 97–98;
    standardization of, 39, 53, 61–63,
    71–78; studies ignoring level of, 143,
    144; theories of, 39, 41–42, 43–46,

60, 68; trends in, 4–6. *See also*
    Curriculum; Funding; Higher
    education; Precollege education;
    Schools; Science/science education;
    Veterinary education
Educational standards. *See* Standards,
    educational
Education Sciences Reform Act of
    2002. *See* No Child Left Behind
Emotions, 17, 25, 46, 134–45
England, 1994 animal use study in,
    121
Engravings, 30
Equality, 63
ERIC database, 186, 196
Ethics, 144
EU. *See* European Union
European Resource Centre for
    Alternatives to Using Animals
    (EURCA) search engine, 50, 184,
    186, 189–90
European Union (EU), 154
Evolution, 61, 96
Experimentation, 49, 84–85, 90;
    phasing out of, 48–49; U.S., quality
    of, 67
Exploratorium, San Francisco, 45–46
Explorit Science Center, 47, 156, 157

Fabricius, Hieronymus, 26
Fabrizzi, Girolamo. *See* Fabricius,
    Hieronymus
Formalin, 102, 121, 159–60, 161;
    toxicity of, 161
Frameworks, standards and, 78–79
France, 1700s, 27
French, R., 19, 20, 28
Frogs, 39, 160, 191, 192
Funding, 3–4, 7, 178–79; alternatives,
    dissection, 51; contention/politics
    and, 60–61; health, 62; legislation
    tied to, 63–64, 96, 99, 178–79;
    science education, 62, 99; standards
    tied to, 178–79; Title I grants, 68, 99;
    veterinary education source of, 51.
    See also Budget

Galen of Pergamum, 18, 22
Games, educational use of, 90

Gender, 146–47; children's animal experiences and, 137; graduation rate and, 66

Google/Google Scholar, 186, 191, 196

Government: legislation levels of, 93; schools run by local, 62; state, 62, 68–69, 100; state v. federal, 62, 68; testing by, 68–69

Grade school: animal use legislation in, lack of, 102; biology/life sciences, 83–87

Graduation rate, high school, 10–11, 66

Graham, Jennifer, 100

Grants, Title I, 68, 99

Grave robbing, 19, 22, 31, 33

Gray's Anatomy, 34

Grove, Mary, Lectures to Ladies on Anatomy and Physiology, 37

Guidelines, precollege schools' optional, 108–9

Guinan, Michael, xii

Hangings, cadavers from, 20, 30, 31–32

Harvey, William, 27

Health: biology and, 10, 93–98, 148–49; families with concerns of, 12–13; formalin exposure and, 102, 121

Health education: goals of, conflicting, 62–63; standards for, 71–78, 82, 87–88; subject matter legislation for, 95–98

Henry VIII, 20

Herbart, Johann Friedrich, 36

Heresy, 26

Higher education, 143, 144; animal use in, 154, 167, 174–76, 178; curriculum design in, 167, 174; interest in, socioeconomics linked to, 67; laboratory animal legislation for, 107–8; precollege v., 2–4, 17, 51, 143, 144, 153–55; socioeconomic slippage in, 5

High school(s): animals dissected in, number of, 153; degrees, U.S. ranking for, 67; dissection adopted by, 119; graduation from, 10–11, 66; laboratories, 89

History, dissection, 3, 18–34, 40; 1300s, 19; 1600s, 27; anatomy study in, 19–20; human bodies in, 17–18

Hogarth, William, 28, 30

Home Office, 154

Home schooling, 12, 46

Hospitals, early history and, 27, 32

The House I Live In (Alcott), 32, 34, 36

Hovey, A., 64

HSUS. See Humane Society of the United States

Human bodies: artistic portrayal of, early history and, 20; high-resolution views of, 91; virtual, 163. See also Cadavers

Human bodies, dissection of, 19, 31, 34, 166–67, 183, 191; animal use linked to, early history, 17–18

Humane Society of the United States (HSUS), 8, 48, 100, 147–53, 162

Hunter, John, 34

Hunter, William, 34

Huxley, Thomas, 38–39

IAHAIO, 109

ILAR. See Institute for Laboratory Animal Research

Individualization, 62

Informal education, 45–46, 143, 157–59

Innovation, 60, 68

Institute for Laboratory Animal Research (ILAR), 102, 108, 186

Instructional Materials Funds, 178

Intelligent design, 61, 96

International Association of Human-Animal Interaction Organizations (IAHAIO), 109

InterNICHE search engine, 13, 50, 184, 187, 190–91

Investigation, 84–85

IQ theory, 63

"Is America flunking Science?", 5

Johns Hopkins University, 191

Jukes, N., 13, 50

Kellert, S.R., 137

Kidd, A.H., 122, 134

Kinzie, M.B., 49, 50, 147
Kits, 160
Klestinec, C., 22, 26
Knowledge: animal, 146–47; artists
    motivating medical educators to
    advanced, 19–20; dissection as new,
    19–20, 21–22, 27, 36, 38; of living
    things, children's, 142–43; scientific
    discoveries and, 36, 38
Knox, Robert, 32
Kozol, Jonathan, 63
Kuhn, T.S., 43, 44, 45

Laboratories, 2, 4, 6, 67, 84–85, 89, 90,
    186; decline in, 51, 88–89; facilities
    for, 114; high school, 89; Huxley
    influence on, 38–39; ineffectiveness
    of, 5–6; legislation for, 107–8;
    resources for, 7, 16; suppliers for,
    181; tradition of, 119. See also
    Experimentation
Language skills, 115
Learner-centric approach, 45–46
Learning: active, 90–91; animal
    structure/function, 136–37; biology,
    10, 145–48; concept, 36, 79, 142–43;
    constructivist, 133–34; family v.
    classroom, 156; learner-centric,
    45–46; "living things," 142–43;
    outcomes, 147; process v. subject
    matter, 89; rote v. meaningful, 43;
    testing v., 7
Lectures to Ladies on Anatomy and
    Physiology (Grove), 37
Legislation, 51, 102–8; animal use, 4,
    93–95, 102–8; biology, 93–98;
    dissection participation, 101–2;
    funding tied to, 63–64, 96, 99,
    178–79; government levels of, 93;
    health education, 93–98;
    laboratories, 107–8; precollege,
    103–8; punishment, 1752, 31; sex
    education, 97–98; subject matter,
    95–98. See also Guidelines
Leonardo. See Da Vinci
Loan libraries, 51, 122, 128, 129–30
Lock, R., 14, 134, 138, 146, 147
London, 34

Mannequins, 176–78
Marcus Aurelius, 18
Marshall, T., 31, 32
Math, 9–10, 66
Matthews, M.R., 43, 44, 111, 133
McInerney, Joseph, 46
Medical schools: alternatives use in,
    176–78; animal use/laboratories in,
    decline of, 88–89; dissection in, early
    history of, 19, 21, 31, 32, 40;
    women's, early history example in,
    40
Mery, Jean, 27
Michelangelo di Lodovico Buonarroti
    Simoni, 20
Middle Ages, 17
Minorities, testing and, 63, 69
Mintzes, J.J., 39, 42, 44
Models, 167
Mondino. See Dei Luzzi
Monitoring, 154
Murray, Charles, 63
Museums, 28, 45–46, 156
Mutilation, 31

NARAL (National Association for the
    Repeal of Abortion Laws)
    Pro-Choice America Foundation, 98
NASCO, 13, 159, 181
National Academies Press, 79, 86
National Anti-Vivisection Society
    (NAVS), 129
National Assessment of Educational
    Progress (NAEP), 68
National Association of Biology
    Teachers (NABT), 108, 179
National Association of State Boards of
    Education, 97
National Center for Education
    Statistics, 5, 68
National Education Association, 39, 41,
    64
National Research Council, 60, 79
National School Boards Association, 87
National Science Foundation (NSF), 42
National Science Teachers Association
    (NSTA), 45, 64, 79, 108, 179
National standards, 71–78, 80–82

"A Nation at Risk: The Imperative for
  Educational Reform," 66
Nation's Report Card, 57, 68
NAVS. *See* National Anti-Vivisection
  Society
NCLB. *See* No Child Left Behind
No Child Left Behind (NCLB), 3,
  63–64, 68, 98–100
NORINA search engine, 13, 50, 184,
  187, 188–89
Novak, Joseph, 44–45
NSF. *See* National Science Foundation
NSTA. *See* National Science Teachers
  Association

Organization for Economic
  Cooperation and Development
  (OECD), 67
Organizations. *See* Animal welfare;
  Professional development;
  Resources; Teaching resources
Orlans, F.B., 14, 107

Parental consent, reproduction study
  needing, 139
Parents/Families, 11–13, 156
Pestalozzi, Johann Heinrich, 36
Pets, 117, 122, 134, 146
Physical principles, in living systems,
  standards for, 84
Physicians Committee for Responsible
  Medicine (PCRM), 102, 153, 162
Physiology, study of, advances in, 20
Piaget, Jean, 43, 142
Pigs, 24, 159
Plastination, 13, 24, 28, 162–63
Plato of Athens, 18
Politics, 60–61, 65, 96
Praying skeleton, 37
Precollege education, 118–20; animal
  use in, 2–4, 17, 103–7, 153–55;
  animal welfare legislation in, 108;
  curriculum design in, 178–79;
  dissection/specimen use in,
  overview, 159–63; higher v., 2–4, 17,
  51, 143, 144, 153–55; legislation for,
  51; standards for health/science in,
  71–78, 82, 87–88

Preservation, 28, 30–34, 161–63;
  plastination type, 13, 24, 28,
  162–63
Professional development, 128, 179–80
Progressivism, pragmatic, 41–42, 60
Public Health Service, 107
Punishment, dissection as
  posthumous, 31

Rabbits, 159
Race, high school graduation
  percentages by, 10–11
Ralph, Charles, 49
Reform, 78–79
Religion, 17, 31, 34
Rembrandt, 28, 29
Renaissance, 19
Reproduction, human, 61, 95, 96, 139
Research literature, teaching resources,
  192–96
Resources, 50–51; legislation, 93–95;
  testing, 16, 54–59; tests replacing,
  problem of, 7. *See also* Teaching
  resources
Resurrection, 17, 31
Richardson, R., 17, 20, 27, 31, 32
Rodents, 154
Rousseau, Jean Jacques, 36
Royal College of Surgeons in London,
  28, 32
Rudolph, J.L., 42, 43, 61

Sachs, Oliver, 138
Salaries, dissector, early, 20
Sappol, M., 28, 32, 34, 36
SAT, 57, 115–16
Sawday, J., 28, 31
Schedules, school: children's, 9–10;
  standards and, 79; teaching, 113–14;
  testing impact on, 64, 116
Schools: grade, 83–87, 102; home, 12,
  46; secondary/intermediate, 118–20,
  122, 155; vocational, 63
Schwab, Joseph, 60
Science Education for Public
  Understanding Program (SEPUP),
  61
Science projects/fairs, 156–57

Science/science education: crisis in, 64;
family role in, 12; goals of,
conflicting, 62–63; informal, 45–46,
143, 157–59; as inquiry, 60, 80–81,
84–85; inquiry goal of, 60; interest
in, 15, 44, 69, 140, 143; introduction
of, 36–39; legislation impact on
funding for, 99; literacy in, 60, 66,
78–79; majors in, 9; national
standards for, 80–81; 1900s, 39–46;
personal/social perspective standards
for, 81; qualifications in, teacher, 99,
117–18; reproduction content in, 95;
Sputnik influence on, 42–43;
standards for health and, 71–78, 82,
87–88; State educational standards
for, 83–87; subject matter in, 83–85,
89; subject matter v. process in, 89;
testing in, 5, 46, 65–67, 68–69; test
scores in, 5, 65–67
Search engines, 50, 183–84, 186–87,
188–91
Season/weather, 28, 31
Segregation, 63
Sentimentality, 46
Servetus, Michael, 26
Sex education, 97–98
Sheep, 157, 160
Siamese twins, 27
SIECUS, 98
Sinclair, Alexander Adam von, 18
Singapore, 5
Singer, S.R., 4, 6, 67, 89
Skinner, B. F., 42
Social issues, 44, 65–67
Social perspectives standards, 81
Socioeconomics, 5, 10, 65, 67
Software, 11, 167, 175
Solutions, preservation, 161–62
SPCA-LA, 135
Species, most commonly dissected,
120–21, 153–55, 159–60
Specimens, 102–8; acquisition of,
159–61; human v. animal, 166–67;
precollege classroom use of, 159–63;
preservation of, 28, 161–63;
suppliers of, 160–61
Spectators, fourteenth century, 19
Spencer, Herbert, 39

Sputnik, 42–43
Standards, educational, 39, 53, 71–78;
educational goals complicating issue
of, 61–63; frameworks and, 78–79;
funding tied to, 178–79; health,
71–78, 82, 87–88; national, 71–78,
80–82; personal perspectives, 81;
science, 71–79, 82, 86–88; state,
82–88; testing and, 65–67, 68–69,
115–16
State government. See Government
Statistics, animal use/dissection, 153,
154
STEM (science, technology,
mathematics), 44
Storm, Bill, xii, 121, 128, 138, 144, 145
Structure/function, in living systems,
84, 136–37
Students: alternative-electing, 51;
Canadian, 14; with disabilities, 69;
education completion level of, 143;
international testing of, 5;
motivating, 8–9, 165; nonmedical,
access for/spectators as, 19; opting
out of dissection, 51, 100, 102,
147–48; science major, 9; unruliness
of, early history and, 22; vivisection
influence on, 18
Studies, 167, 183; animal
structure/function learning, 136–37;
animal use in dissection, 120–21;
attitudes toward animals, 143;
conversational analysis of children,
136; dissection v. alternatives, 147;
educational use of dissection/dead
animals, 145; learning outcome, 147;
student age/education level in, 143,
144; veterinary method comparison,
50; whole v. parts of animals, 138–39
Subject matter, 113, 114;
biology/health, 83–85, 95–98;
dissection-related, 116; process v., 89
Sudan, Africa, 40
Suppliers: laboratory, 181; specimen,
160–61; teaching resources, 180–82;
textbook, 180–81
Surgeons, history of dissection and, 19,
28, 32
Surrogates, animals as, 18, 183

Teachers, 6–9, 90, 111; attitudes/practices of, 120–21; constrictions on, 61–62, 95; consultants for, 196; curriculum design by, 174–75; dissection viewed by, 14, 49; elementary/secondary, experiences of, 122; first, 19; job satisfaction of, 120–21; professional development for, 128, 179–80; science qualifications of, 99, 117–18; student motivation concern of, 8–9; uninvolvement of, 49

Teaching: anatomy, 11; animals used in, information on, 153–55; challenges in, 113; innovative, 68; methods, 90–91; preparation/rewards in, 112–13; schedules, 113–14; subject matter in, 83–85, 89, 95–98, 113, 114, 116

Teaching resources, 16, 116, 183–97; alternatives, search engines for, 50, 122–28, 183–84, 186–87, 188–91; animal use guidelines, 108–9; biology, 166, 168–74; budget for, 116, 119; database, 185–91; lack of, 6–8, 119; loan library, 51, 122, 128, 129–30; research literature, 192–96. *See also* Suppliers

Technology, 44, 114; availability of, 10, 13–14, 130–31, 163; computer, 11, 90, 91, 125, 131, 147, 163, 167, 175. *See also* Preservation

Terman, Louis, 42

Testing, 59–65; animals used in, 153–55; home schooling and, 46; international, 5; learning v., 7; norm-based v. criterion-based, 64; pressure of, 115–16; resources for, 16, 54–59; science, 5, 46, 65–67, 68–69; standardized, 65–67, 68–69, 115–16; time spent, increased, 64, 116

Texley, Juliana, 99

Textbooks, 68, 71, 178, 179; review of, 68; suppliers of, 180–81

Theatres, 27–28

Thorndike, Edward, 42

Tobacco, 96–97

Trends in International Mathematics and Science Study (TIMMS), 4, 5, 67

Tufts University, 50

Tunnicliffe, S.D., 12, 136, 137, 143, 156

UC Center for Animal Alternatives, 50

Umbilical cord, 27

United Kingdom, 154

United States (U.S.): high school degrees, ranking of, 67; HSUS of, 8, 147; laboratory experiences rating for, 67

University of California, Davis, School of Veterinary Medicine, xii, 175–76

University of California Office of the President, 82, 88

U.S. Department of Agriculture, 102, 107

U.S. Department of Education, 4, 58, 59, 63, 64, 67, 68

Vertebrate animals, 153–54

Vesalius, Andreas, 18, 20, 22, 23, 24, 26, 28; conviction of, 34; *De Humani Corporis Fabrica (On the Construction of the Human Body)*, 22, 23, 25

Veterinary education, 50, 174–75; alternatives introduction into, 175–78; consumptive animal use stopped in, 175; phasing out in, 2, 14, 50, 51

Visible Human, 91

Vivisection, 18, 24, 129

Vocational schools, 63

Voltaire, 26

Von Hagens, G., 13, 24

WARD's Natural Science, 159, 160, 162, 181

Wax figures, 28

Women's medical college, 40

Writing skills, 9–10

Zasloff, R.L., 102, 117, 135, 146

Zirkel, J.B., 50, 147

Zoos/zoology, 136, 143, 156, 158, 159

About the Authors

LYNETTE A. HART is professor in the Department of Population Health and Reproduction in the School of Veterinary Medicine at the University of California, Davis.

MARY W. WOOD is Librarian in Health Sciences at the Carlson Health Sciences Library, University of California, Davis.

BENJAMIN L. HART is Distinguished Professor Emeritus of Behavior and Physiology in the Department of Anatomy, Physiology, and Cell Biology in the School of Veterinary Medicine at the University of California, Davis.